Fluid Power
Educational
Series

Pneumatic Systems and Circuits -Basic Level

(In the SI Units)

Joji Parambath

Pneumatic Systems and Circuits – Basic Level
(In the SI Units)

ISBN: 9798651963027

https://jojibooks.com

First Edition: 2020
Revised Edition: 2021
Revised Edition: 2022
Revised Edition: 2023

Disclaimer of Liability

The contents of this textbook have been checked for accuracy. Since deviations cannot be avoided entirely, we cannot guarantee full agreement. Only qualified personnel should be allowed to install and work on pneumatic equipment. Qualified persons are those authorised to commission, ground, and tag circuits, equipment, and systems following established safety practices and standards.

Dedicated to

my loving mother, Padmavathy

Table of Contents

Preface

'Pneumatic Systems and Circuits - Basic Level (in the SI Units)' is an introductory textbook dealing with pneumatic system components and circuits. The fundamentals required to understand the core topics are given initially. The book describes the topics of compressed air generation and contamination control, pneumatic actuators, and control valves, in detail. Further, the book presents the maintenance, troubleshooting, and safety aspects of pneumatic systems. The textbook uses the SI system of units.

The topics are presented in a logical sequence and simple to understand language. Many critical positions of the circuits are given wherever possible to make the reader understand the control circuits easily. Many exercises are given at the end of the chapters to check their understanding of the subject.

Many other fluid power topics are given in other textbooks under the fluid power educational series by the same author. A list of all the books is given at the end of the book. Also, please see the details at: https://jojibooks.com.

Enjoy reading the book.
Your feedback is most welcome.

JOJI Parambath

About the Author ….

Joji Parambath is a trainer in the field of Pneumatics, Hydraulics, and PLC, for over 25 years. During his career, he has trained numerous professionals from the industries as well as faculty members and students of engineering institutions.

At present, he is the key trainer at Fluidsys Training Centre, Bangalore, India, (https://fluidsys.org), which provides training in Pneumatics and Hydraulics. He has already written two books on Pneumatics and Hydraulics. The publication of the present series of 36 books is intended to restructure and update the existing books.

The author wishes to thank all trainees for their lively interaction and many useful suggestions during the training programmes that prompted the author to write the present series of books. You may send your feedback to joji.p@hotmail.com

10th June 2020

A Note on the Revised Edition 2023 ….

The third revised edition of the book has been prepared, eliminating typographical errors in previous editions and incorporating more useful information to make the book more reader-friendly and valuable.

JOJI Parambath

Chapter 1 | Industrial Power Systems

Modern industrial production systems and mobile systems are designed to carry out a wide variety of work operations. A prime mover provides the muscle power for performing a work operation in a stationary production machine or mobile equipment. The prime mover is, in fact, an actuator that is part of a power transmission system consisting of a power source and a control system. Remember, the power must be conveyed to the load at the machine's point of work through the power transmission system in a controlled manner. The block diagram of an industrial power system is given in Figure 1.1.

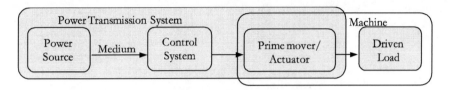

Figure 1.1 | Block diagram of an industrial power system

Apart from the cumbersome mechanical linkages, power can easily be transmitted through the electron medium or air medium or oil medium. Accordingly, there are three main types of power transmission systems. They are: (1) Electrical power transmission system, (2) Pneumatic power transmission system, and (3) Hydraulic power transmission system.

Next, these power transmission systems are extensively used in modern industrial systems and mobile equipment. Today's manufacturing and material handling systems demand huge power, compact machines, faster operations, better quality, greater efficiency, and lower cost. At the same time, power transmission systems have grown superbly to meet the requirements of engineering systems. Moreover, they are also found to be quite amenable to automation and lend themselves to design simplification. That means these power transmission systems will remain the backbone of industrial machines and mobile equipment, and continue to expand in times to come.

Electrical Power Transmission System

The block diagram of an electrical power system is given in Figure 1.2. An electric generator produces electrical power. The power is then transmitted, in the form of electron flow, to loads, such as electric motors, lamps, heaters, etc., through a final control element, such as a contactor. A relay controller or PLC controller controls the final control elements, which, in turn, control the loads.

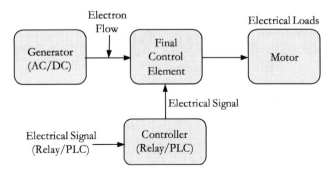

Figure 1.2 | Block diagram of an electrical power system

Pneumatic Power Transmission System

The block diagram of a pneumatic power system is given in Figure 1.3. Compressed air is generated by a compressor when driven by its prime mover, such as an electric motor. The energy in the form of compressed air is then transmitted to actuators, such as cylinders, pneumatic motors, etc., through final control elements, such as directional control valves. A pneumatic controller, relay controller, or PLC controller controls the final control elements, which, in turn, control the pneumatic actuators.

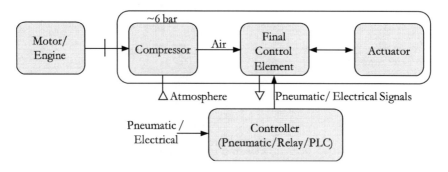

Figure 1.3 | Block diagram of a pneumatic power system

Pneumatics is the engineering science of gaseous pressure and flow. Pneumatic systems are low pressure systems. They are especially suitable for systems that involve low forces or high-speed linear motions or both.

Hydraulic Power Transmission System

The block diagram of a hydraulic power system is given in Figure 1.4. Pressurised oil medium is generated by a pump driven by a prime mover, such as an electric motor. The energy in the form of pressurised oil is then transmitted to actuators, such as cylinders, hydraulic motors, etc., through final control elements, such as directional control valves. A relay controller or PLC controller controls the final control elements, which, in turn, control the hydraulic actuators.

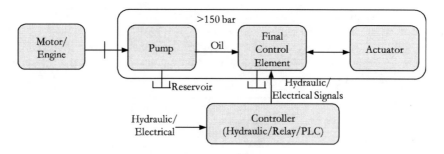

Figure 1.4 | Block diagram of a hydraulic power system

Hydraulics is an engineering science of liquid pressure and flow. Hydraulic systems are high-pressure systems. They are suitable for systems that require precise slow-speed control or involve the holding of heavy loads.

Fluid Power Transmission System

Pneumatic and hydraulic power systems are commonly categorised under fluid power systems.

Significant Criteria for the Selection of Power Transmission Systems

Electrical motors are the optimum devices for obtaining rotary motions. Hence, they are used in systems involving predominantly rotary motions. Pneumatic power is suitable for systems that involve high-speed linear motions, but low forces. However, a major drawback of pneumatic systems is that they are not suitable for obtaining uniform motions. Hydraulic power is suitable for systems that require very smooth position control or demand holding of heavy loads.

Evolution of Fluid Power Transmission Systems

Fluid power systems evolved from manual to mechanisation to automation. In mechanisation, the mechanical work process is taken over by a machine. In automation, a machine is controlled automatically with limited human intervention or without human intervention at all.

Control System

The control part in a power transmission system modulates the power part using control elements generally known as 'final control elements' to control work processes. A final control element can be operated directly by a manual force or through pneumatic or electrical signals produced by some input elements such as pushbuttons and sensors (Figure 1.5). The final control element can also be operated indirectly through a pneumatic or relay or electronic or PLC controller.

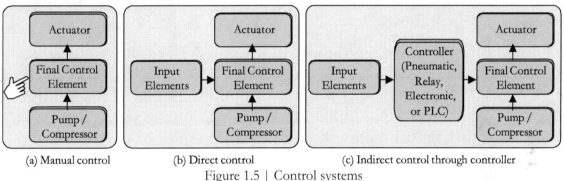

(a) Manual control   (b) Direct control   (c) Indirect control through controller

Figure 1.5 | Control systems

Concept of Automation

The work process to carry out an industrial task usually involves many recurring steps. These steps can be carried out manually or automatically. In the manual system, an operator is always present to decide every process step. In the automatic system, the process controls itself, partially or entirely, by the feedback of its condition. Therefore, an automatic system can be semi-automatic or fully automatic. In the semi-automation, a machine automatically carries out several recurring partial steps in the processing of a work-piece. Here an operator is required to initiate every cycle of operations. Whereas, in a complete automation process, a machine carries out the cyclic operations in the processing of several jobs automatically. Sensors and/or transducers are invariably used in automatic control systems.

Motion Control Systems

The motion control system is an extension of modern automation systems. It mainly controls the position, velocity, and force associated with work operations. A motion controller is the brain within the motion control system. It is responsible for calculating and generating the output commands for the desired motion path or trajectory. Motion control is an intricate part of a robot or modern CNC machine.

Comparison of Different Power Transmission Systems

Choosing the right and efficient form of energy for the drive system in the industry is not an easy task. Its selection depends on various factors. Table 1.1 gives a comparison of different forms of energy medium based on some essential criteria, as mentioned.

Table 1.1 | Comparison of different power transmission systems

Criteria / Power system	Electrical	Hydraulics	Pneumatics
Energy production	Hydro, thermal, atomic	Pump	Compressor
Availability of energy transmission medium	Available everywhere	Obtaining and disposing of oil is expensive	Air is freely available
Maximum distance for energy transmission	Considerable distance, even beyond 1000 km	Up to 100 m	Up to 1000 m
Cost of energy	Smallest	High	Highest
Speed control	Limited	Good for slow speed precise control	Best for high-speed operation, obtaining uniform speed difficult
Linear force	Using rotary to linear conversion devices	Using cylinders - Large forces due to high pressure	Using cylinders - Limited forces due to low pressure
Rotary force (Torque)	Using electric motors	Using hydraulic motors	Using air motors
Overloading	A severe problem	loadable until standstill with relief valves	Loadable until standstill
Sensitivity to temperature variations	Insensitive	Sensitive	Relatively insensitive
Leakage	Lethal accident risk at high voltages	Loss of energy and environmental fouling	Loss of energy

Objective-type Questions

1. The large magnitude of linear forces can be obtained easily in:
    a) Mechanical power transmission systems.
    b) Electrical power transmission systems.
    c) Pneumatic power transmission systems.
    d) Hydraulic power transmission systems.

2. Which of the following power transmission systems does provide a fast-acting production system?
    a) Mechanical power transmission system.
    b) Electrical power transmission system.
    c) Pneumatic power transmission system.
    d) Hydraulic power transmission system.

3. Which of the following statements is <u>incorrect</u>?
   a) Pneumatic systems are overload-safe.
   b) Hydraulic systems are insensitive to variations in temperature.
   c) Pneumatic systems are capable of providing high-speed operation.
   d) Hydraulic energy can be transmitted economically, typically up to 100 m.

4. Which of the following statements is <u>correct</u>?
   a) The electrical power system provides linear motions in an optimum manner.
   b) The pneumatic power system provides uniform motion of its actuators.
   c) The hydraulic power system is not suitable for getting rotary motions.
   d) A motion control system calculates and generates output commands for the desired trajectory of motion.

5. The function of a controller in a power system is to:
   a) transmit power through the system.
   b) regulate the pressure in the system.
   c) govern the primary power system through commands.
   d) sense the output parameter of the system.

Questions
1. What is an industrial prime-mover?
2. What are the essential components of industrial power transmission systems?
3. What is the primary function of power transmission systems?
4. What is a fluid power system? Explain briefly.
5. What are the main divisions of fluid power systems?
6. List some essential basic functions performed by fluid power systems.
7. What is the major advantage of fluid power systems?
8. What are the drawbacks of fluid power systems?
9. List any four applications of fluid power systems.
10. List two applications of oil hydraulics.
11. Describe some unique problems faced by fluid power systems.
12. Compare hydraulic and pneumatic systems.
13. Explain why a control system is required in power transmission systems.
14. Briefly describe the evolution of industrial work processes.
15. What are mechanisation and automation?
16. Differentiate between 'semi-automation' and 'complete automation'.
17. Give one example each of 'semi-automatic control' and 'fully automatic control'.
18. What is a motion control system? Explain briefly.
19. Mention three advantages of hydraulic systems as compared to other power systems.
20. Draw the essential blocks of the pneumatic energy transmission system and explain.
21. Depict the essential elements of the hydraulic energy transmission system with the help of a block diagram and describe the primary function of each element.

Answer key for objective-type questions:
Chapter 1: 1-d, 2-c, 3-b, 4-d, 5-c

Chapter 2 | Pneumatic Fundamentals

Pneumatics is the branch of engineering sciences concerned with the transmission of energy using compressible fluids, like air, etc. Pneumatics is used throughout the industry due to the versatility and simplicity of its application. Many characteristics make pneumatics more appropriate for many industrial applications, as compared to other types of power transmission systems.

How do we Define Pneumatics?

The term pneumatics is derived from the Greek word *'**pneuma**'*, which means wind or breath. Therefore, the study of pneumatics deals with systems operated with compressed air medium to impart and control power. The study of pneumatics is all about knowing how to produce a positive pressure using a force through the medium of compressed air, and the reverse process of how to develop force and control a load, by making use of the developed pressure.

Energy Transfer Medium in Pneumatics

In industrial pneumatic systems, the most commonly used medium for transmitting power is highly compressible air. Some systems use nitrogen and natural gas as power transmission media.

Composition of atmospheric air

Dry air at sea level is composed of 78% nitrogen, 21% oxygen, and 0.93% argon by volume. It also contains traces of carbon dioxide, hydrogen, neon, helium, krypton, and xenon. (See Figure 2.1) However, a pneumatic professional needs to note that atmospheric air holds many harmful impurities, like dust, water vapour, oil particles, etc.

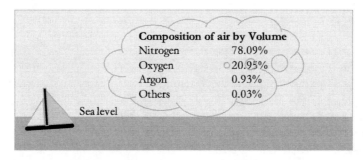

Figure 2.1 | Composition of atmospheric air

Gas laws

Air is a mixture of gases and follows the laws of perfect gas concerning its behaviour in volumetric expansion or contraction and absorbing or releasing heat. The air medium is sensitive to changes in volume, pressure, and temperature.

Boyle's law

Boyle's law gives the relation between the pressure and volume of a gas. It states that: 'At constant temperature, the volume of a given mass of gas is inversely proportional to its absolute pressure.' An illustration of Boyle's law is given in Figure 2.2. Let V_1 (4 l) be the volume of a gas at pressure P_1 (1 bar). When the gas is compressed to a volume V_2 (2 l), then the pressure will rise to a value of P_2 (2 bar).

Mathematically,

$$P_1V_1 = P_2V_2, \text{ where the temperature remains constant}$$

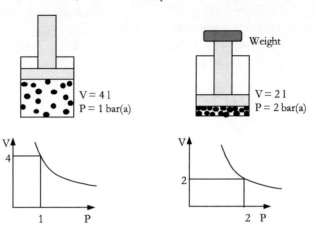

Figure 2.2 | Illustrating Boyle's law

As air is compressed, the energy used in this work is dissipated as heat. That is, the temperature of the compressed air will rise, as the intake air is reduced in volume.

Gay- Lussac's Law

The law states that: 'If the volume of a given mass of gas is held constant, the absolute pressure of the gas varies directly to its absolute temperature.'

$$\frac{P1}{T1} = \frac{P2}{T2}; \text{ V remains constant}$$

Combined Gas Laws

The general law explains how the variables of absolute pressure, volume, and temperature are related to each other in a fixed mass of gas. The law can be expressed mathematically as:

$$P1 \cdot \frac{V1}{T1} = P2 \cdot \frac{V2}{T2}$$

Air Compression Process

The compression (or expansion) of air takes place under isothermal, adiabatic, or polytropic conditions.

Isothermal Compression

If the compression of air takes place under constant temperature conditions, the process is said to be isothermal. That means the heat of compression must be removed at the same rate as it is produced. Therefore, the process must be slow enough for the heat of compression to flow out of the air, as it is compressed. The equation governing the isothermal compression can be stated mathematically as:

PV is a constant

However, in practice, it is not possible to take out all the heat as it is generated.

Adiabatic Compression

When a volume of air in a system is compressed (or expanded) instantly, there is no time to add (or dissipate) heat into (or out of) the system, and this type of compression process is said to be adiabatic. For example, adiabatic compression takes place when air is compressed in a fully-insulated cylinder without any possibility of heat exchange with the surroundings. The same is the case with the air expanding through a nozzle very quickly. The equation governing the adiabatic compression can be stated mathematically as:

$$P V^n \text{ is a constant}$$

The value of n for the adiabatic compression of air is taken as 1.4.

Polytropic Compression

An isothermal compression process must occur very slowly to keep the air temperature constant. An adiabatic compression process must occur very rapidly without any flow of energy into or out of the system. These compression processes are considered to be theoretical and hence are presumed to be taking place under ideal conditions. In actual practice, compression of air occurs between the two limits of compression. The polytropic compression process represents the actual compression process in compressors and pneumatic actuators operating under the normal rate of compression and expansion. For the polytropic compression:

$$P V^n \text{ is a constant}$$

The value of n for a poytropic compression depends on the type of gas and the rate of compression and is less than 1.4. Typically, for air, the value of n is taken as 1.3.

Characteristic Curves for the Compression Processes

The characteristic curves for the isothermal, adiabatic and polytropic compression processes are given in Figure 2.3.

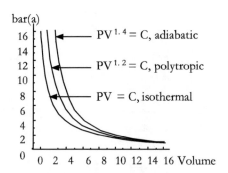

Figure 2.3 Characteristics of compression process for air

Pascal's Law

Pascal's law is central to the development of many fluid power devices, such as brakes, presses, and jacks. The law can be stated in two parts: (1) 'Pressure at any one point in a static fluid is the same in every direction' (See Figure 2.4). (2) 'Pressure exerted on a confined fluid is transmitted equally in all directions, acting with equal force on equal areas'.

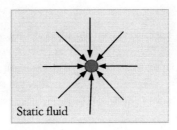

Static fluid

Figure 2.4 | Pascal's law

Pneumatic Pressure

Pressure in pneumatics operates according to Pascal's law. Thus, pressure is the distributed response of force acting through a fluid.

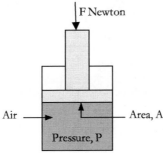

F Newton

Air →   Area, A

Pressure, P

Figure 2.5 | Pressure development in confined air

In Figure 2.5, a definite amount of force (F) is applied to the air enclosed in the chamber, using a piston with an area A. The enclosed air is compressed, and its pressure (P) rises in direct proportion to the applied force and is inversely proportional to the area of the piston. Pressure can, therefore, be defined as the force acting per unit area.

$$P = \frac{F}{A}$$

Units of Pressure

In the SI system, the unit of pressure is Pascal (Pa), and 1 Pascal is the constant pressure acting on a surface area of 1 square metre with a perpendicular force of 1 Newton. For industrial pneumatic purposes, Pascal is too small a unit for use in measurements, and hence more practical units like bar, kilo Pascal, mega Pascal, etc. are used.

1 Pascal	$= 1 \text{ N/m}^2$	
1 bar	$= 100000$ Pa	$= 10^5$ Pascal
1 Mega Pascal (MPa)	$= 10^6$ Pascal	$= 10$ bar
1 Kilo Pascal	$= 10^3$ Pascal	
1 bar	$= 0.1$ MPa	
1 bar	$= 14.5$ Pound per square inch (psi) [lb/in²]	
1 bar	$= 1.02 \text{ kgf/cm}^2$	
1 kgf/cm²	$= 0.981$ bar	

Pressure Scales

Everything on the earth's surface is subjected to a significant pressure head from the weight of the air above. This pressure is the 'atmosphere' (atm) and is approximately equal to 1 bar (more precisely, 1.01325 bar) at sea level. A pressure gauge is capable of measuring only the pressure with reference to the local atmosphere. Therefore, the measured value of the pressure does not include the pressure exerted by the atmosphere. However, we require pressure values with reference to the absolute vacuum, especially for use in calculations. Therefore, two pressure scales are specified in pneumatic systems according to the reference pressure levels. They are: (1) Gauge pressure scale and (2) Absolute pressure scale.

Gauge Scale

The gauge pressure is the pressure indicated by a pressure gauge installed at a particular location. It is the pressure above the local atmospheric pressure, regardless of the altitude. The gauge pressure, measured in bar, can be stated as bar(g) or simply bar. The gauge pressure, measured in psi, can be stated as psi(g) or simply psi.

Absolute Scale

The absolute pressure scale begins at the point where there is a complete vacuum (zero absolute pressure). The absolute pressure value can be obtained by adding the datum pressure level [for example, 1.013 bar or 14.7 psi at sea level] to the gauge pressure level. The absolute pressure measured in bar should be stated as bar(a). The absolute pressure measured in psi should be stated as psi(a).

The relationship between the absolute pressure and the gauge pressure is illustrated graphically in Figure 2.6. Zero gauge pressure indicates the local atmospheric pressure (absolute).

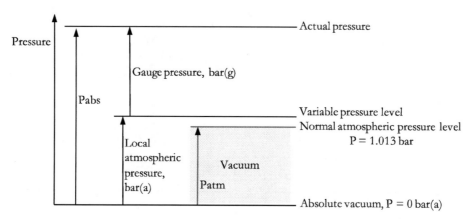

Figure 2.6: Pressure scales

Economic Pressure in Pneumatic Systems

Pneumatic systems have been developed and progressed comparatively as low-pressure systems, as the compression of air is found to be a slow process. Pneumatic air-consuming devices such as cylinders and air motors are generally designed for a maximum operating pressure of 8 to 10 bar. However, practical experience has shown that 6 bar is the ideal pressure for the economic operation of pneumatic systems. This low pressure allows the designer to keep the size of pneumatic components very compact and maintain the cost of the components and piping system to a minimum.

Industrial Pressure Ranges

In most industrial pneumatic systems, the preferred operating pressure range is from 6 to 10 bar. Many popular air tools are engineered for pressures between 6 and 7 bar. However, the extended pressure range for industrial pneumatic systems can be up to 16 bar. Control pressures in pneumatics can be as low as 3 bar. The industrial pressure ranges in absolute and gauge pressure scales are shown in Figure 2.7.

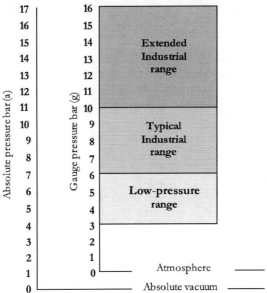

Figure 2.7 | Industrial pressure ranges

Problem 2.1

1120 litres of air at atmospheric pressure is compressed to 160 litres. What will be the gauge pressure developed, if the temperature remains the same?

Solution

Initial volume (V1)	= 1120 litre
Initial pressure (P1)	= 1 bar(a)
Final volume (V2)	= 160 litre
Final pressure (P2)	= V1 x P1 / V2
	= 1120 x 1 / 160 = 7 bar(a) = 6 bar(g)

Problem 2.2

Calculate the pressure produced by a force of 10000 N acting on the piston with an area of 0.05 m².

Solution

Force	= 10000 N
Area	= 0.05 m²
Pressure	= F/A
	=10000 / 0.05 = 200000 Pa = 2 bar

Pneumatic Force

Let us now understand the process of developing a force to drive a load in the pneumatic system by the application of pressure. Figure 2.8 shows the schematic diagram of a pneumatic cylinder with a piston. When the pressure (P) is applied to the area (A) of the piston, it develops a force (F). The amount of force developed is equal to the applied pressure times the area.

That is, $F = P \times A$

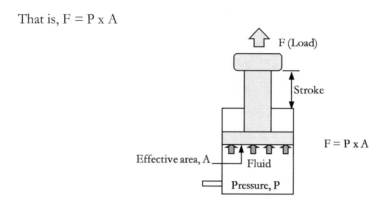

Figure 2.8 | A cylinder developing a force (F) with the application of the pressure (P)

Force Multiplication

A pneumatic system can be designed for easy force multiplication. Figure 2.9 shows an arrangement of two cylinders A and B with piston areas A1 (say 1 sq cm) and A2 (say 10 sq cm) (A2 > A1) respectively interconnected through a pipe. The enclosed space, inside the cylinders and the pipeline, is filled with air. When cylinder A is applied with a force F1 (say 10 N), a pressure P (10 N/sq cm) is generated in the air medium. The same pressure P acts on the piston of cylinder B, as per Pascal's law. This pressure causes the development of force F2 (100 N) by cylinder B. The governing equations for the forces applied or developed in the cylinders are as follows:

$$F1 = P \times A1$$
$$F2 = P \times A2$$

Therefore,

$$F2 = F1 \times \frac{A2}{A1}$$

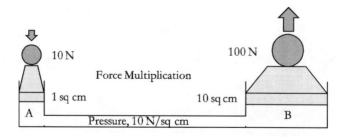

Figure 2.9 | Illustration of force multiplication concept

12

Thus, a pneumatic system can be designed for force multiplication. The ability of pneumatic systems to realise force multiplication can be thought of as leverage. However, it may be noted that force multiplication is achieved by sacrificing distance. That is, for example, if cylinder A moves by 10 cm, then cylinder B moves only by 1 cm.

The Flow Rate of Air
The flow rate of air is the volume of air passing a cross-section per unit of time under the specified conditions of pressure, temperature, and relative humidity. It is usually measured in terms of lpm or cfm.

Problem 2.3
What force is produced by a pneumatic cylinder with an area of 20 cm^2 operating at a pressure of 6 bar?

Solution

Area, A	$= 20$ cm^2 $= 0.002$ m^2
Pressure, P	$= 6$ bar $= 6$ x 10^5 Pa
Force, F	$=$ P x A
	$= 6$ x 10^5 x $0.002 = 1200$ N

Problem 2.4
A pneumatic lift system, consisting of a small cylinder of bore diameter 32 mm and a large cylinder of bore diameter 200 mm, is to lift a load of 4000 N. What is the force required to be exerted on the piston of the small cylinder to lift the load?

Solution

The bore diameter of the large cylinder	$= 200$ mm
The bore diameter of the small cylinder	$= 32$ mm
Force to be lifted, F2 $= 4000$ N	
The piston area of the small cylinder, A1	$= \prod . 32^2 / 4$
	$= 803.84$ mm^2
The piston area of the large cylinder, A2	$= \prod . 200^2 / 4$
	$= 31400$ mm^2
Therefore,	
Force to be exerted to cylinder A, F1	$=$ F2 . (A1 / A2)
	$= 4000$ x $(803.84 / 31400)$
	$= 102.4$ N

Absolute Temperature
Calculations involving volume and pressure changes of air must be performed using absolute pressure and temperature values. The absolute temperature (K) [degree Kelvin] in the SI system of units is determined as the sum of the measured temperature (°C) and 273. That is,

Absolute temperature (K) = Measured temperature (°C) + 273

Objective-type Questions

1. Boyle's law on ideal gases states that:
   a) The air compression process is always isothermal
   b) Air medium can be used for force multiplication
   c) The pressure at any one point in a static fluid is the same in every direction
   d) Pressure increases when volume decreases, under a constant temperature condition

2. If air is compressed at a constant temperature, the compression process is:
   a) Adiabatic
   b) Isobaric
   c) Isothermal
   d) Polytropic

3. If air is compressed without the addition of heat or the removal of heat, the compression process is:
   a) Adiabatic
   b) Isobaric
   c) Isothermal
   d) Polytropic

4. The actual air compression process is regarded as:
   a) Adiabatic
   b) Isobaric
   c) Isothermal
   d) Polytropic

5. 6 bar is the same as:
   a) 0.6 Kgf/cm^2
   b) 600 kPa
   c) 6 MPa
   d) 870 psi

6. 100 psi is equal to:
   a) 0.6 bar
   b) 0.6 Kgf/cm^2
   c) 6 MPa
   d) 6.9 bar

7. 600 kPa is the same as:
   a) 0.6 Kgf/cm^2
   b) 60 bar
   c) 0.6 MPa
   d) 8.7 psi

8. The most economic pressure in pneumatics is
   a) 1 bar
   b) 6 bar
   c) 6 psi
   d) 90 bar

9. Gauge pressure is
   a) 1 bar less than the absolute pressure
   b) 1 bar more than the absolute pressure
   c) Measured with reference to the absolute vacuum
   d) The same as the absolute pressure

10. A compression process that takes place too slowly is typically:
    a) Adiabatic
    b) Isobaric
    c) Isothermal
    d) Polytropic

11. A compression process that takes place too fast is typically:
    a) Adiabatic
    b) Isobaric
    c) Isothermal
    d) Polytropic

Review Questions
1. Explain briefly how an air medium can be used for transmitting power.
2. Explain Boyle's law.
3. How are the variables pressure, temperature, and volume related in the case of gas?
4. Explain the nature of the true compression process in a pneumatic system concerning the addition or release of heat.
5. Differentiate: Adiabatic compression process and isothermal compression process.
6. State and explain Pascal's law.
7. Briefly explain how pressure is developed in pneumatic systems.
8. Name the two pressure scales used for measurement purposes and differentiate between them.
9. Why is 6 bar regarded as the most economic pressure in pneumatics?
10. Give typical values of the pressure ranges for industrial pneumatic applications.
11. A mass of 500 Kg needs to be pushed upwards using a double-acting pneumatic cylinder. What diameter of cylinder do we need if the pressure available is 6 bar? Ans: 100 mm
12. Determine the bore size of a pneumatic cylinder to keep a work-piece pressed with a static clamping force (F) of 1000 N during the extension stroke and at the operating pressure (P) of 5 bar. Consider the load factor as 0.7. Note: The actual output force produced by the cylinder is lower than its theoretical force output (F=P x A) by the load factor due to frictional and sliding resistances.
    {Hint: Find (1) Theoretical force, cylinder = Clamping force / Load factor, (2) Area of the cylinder piston, A (m²)=F (N)/P (Pa), and (3) Bore diameter, D=√ [(4 x A)/π]} Ans: 60 mm
    [Solutions to problems 11 and 12 are given at the end of the book]
Answer key for objective-type questions:
Chapter 2: 1-d, 2-c, 3-a, 4-d, 5-b, 6-d, 7-c, 8-b, 9-a, 10-c, 11-a

Chapter 3 | Advantages and Disadvantages of Compressed Air Systems

Pneumatic systems are developed as low-pressure systems using compressed air to drive small loads. The following sections highlight the advantages and disadvantages of pneumatic systems.

Advantages of Compressed Air Systems

A compressed air system is a versatile and straightforward method of transmitting energy. The use of a compressed air system may allow the simplification of machine design. Components of air systems are usually very compact and lightweight and can be easily installed and serviced. There are fewer mechanical parts in compressed air systems. Hence, pneumatic systems are more efficient and more dependable. Air may be exhausted to the atmosphere without any harm, so return lines are not required in pneumatic systems. Compressed air systems may be used in hazardous areas where electrical controls cannot be employed. A few of the most important positive characteristics of compressed air are given below.

Quantity

Air is available everywhere in unlimited quantities

Power Transmission

Compressed air can be easily transmitted through pipes over long distances, up to 1000 m.

Storage

A considerable quantity of compressed air can be easily stored in a receiver tank. This charging of the receiver tank is the process of storing potential energy. The system can receive the compressed air directly from the tank, and the compressor need not be in operation always.

Speed

Compressed air is a fast-working medium. The operating speeds of pneumatic cylinders can be as high as 2 m/s.

Speed Control

Actuator speeds can easily be controlled by using simple valves.

Acceleration

Air is extremely compressible and elastic and is capable of absorbing large amounts of potential energy. These properties make possible the use of compressed air to obtain quick acceleration and deceleration of actuators and reversal of direction of motions with relative freedom from shock.

Control

Since force can be easily controlled, mechanical elements driven by compressed air systems can be stalled for infinite periods without any damage. A compressed air system can quickly and efficiently be controlled with a few control elements and can readily be adapted for automation.

Overload

Compressed air tools and working elements can be subjected to loads even at a standstill and are, therefore, overload safe.

Disadvantages of Compressed Air Systems

Although the advantages of compressed air systems are numerous, they are counteracted by certain disadvantages. To be able to establish a clear demarcation in the field of pneumatic applications, it is necessary to become familiarised with the negative characteristics of compressed air systems as given below.

Preparation

All compressed air systems are vulnerable to damage by solid contaminants, moisture, and oil particles. Hence, compressed air requires proper preparation to remove undesirable elements present in it and achieve the desired cleanliness level.

Force Limitation

Compressed air actuators are economical only up to a specific force requirement. Based on the normal operating pressure of 6 bar, the limit is approximately 50 kN.

Analytical Complexity

Pneumatic systems are complicated due to the high compressibility and nonlinearities of the flow characteristics.

Uniform Speed

It is not possible to achieve uniform and constant piston speeds with compressed air. Compressibility and elasticity of air can impair operations unless these characteristics are correctly understood and used.

Heat Loss

As air is compressed, it becomes very hot. This heat generation represents a loss of energy.

Costs

Compressed air is a relatively expensive energy medium. However, inexpensive components and high performance compensate for the high costs of energy.

Exhaust Air

Exhaust air is noisy. Nowadays, this problem has been largely solved with the availability of effective silencers

Objective-type Question

1. Which of the following statements is correct?
   a) Pneumatic actuators cannot withstand overloads.
   b) Pneumatic systems are superb for obtaining uniform speeds.
   c) Pneumatic energy can be transmitted only up to a distance of 50 m.
   d) Compressed air systems are vulnerable to heat, dirt, moisture, and oil particles.

Review Questions

1. List out four essential advantages of compressed air systems.
2. List out three disadvantages of compressed air systems.
3. Give two advantages of systems using compressed air concerning the speed of system actuators.
4. Elaborate on the following disadvantages of compressed air in a pneumatic system:
   (a) Contamination and (b) Force limitation.

Chapter 4 | Compressed Air Generation and Storage

The power source in a pneumatic system must be designed to supply a sufficient quantity of compressed air at the required pressure to all actuators in the system. The primary functions of the power source include the generation and storage of compressed air, regulation of pressure, removal of unwanted heat and contaminants from the compressed air, and the distribution of the compressed air. Accordingly, the main components of a pneumatic power source include a compressor, a receiver tank, pressure regulators, a cooler, filters, and dryers. A schematic diagram of a typical pneumatic power source, with elements for compressed air generation, storage, and preparation, is shown in Figure 4.1.

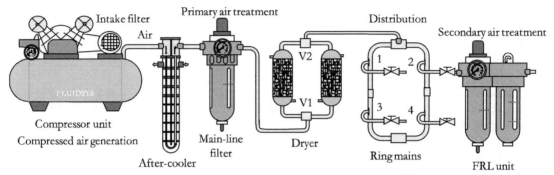

Figure 4.1 | Compressed air generation and storage

A compressor is designed to take in air at atmospheric pressure and deliver it into a closed system at a higher pressure for generating forces needed to perform some useful tasks. The pressure in the system is regulated and limited using pressure-regulating units. The compressed air preparation elements, such as coolers, filters, and dryers, remove undesirable elements, like heat, solid contaminants, moisture, and oil particles present in the compressed air. Note, the preparation of compressed air takes place in various stages. A distribution system is intended to supply the compressed air uniformly to all actuators in the system. In short, the power source must be able to supply clean and dry compressed air to all the consuming devices in the system at the required pressure and in sufficient quantity. Further, the temperature must be maintained within limits and leakages should be minimized as far as possible. This chapter presents the topics of compressed air generation and storage.

Air Compressors

Compressors are the most common industrial energy supply units in pneumatic systems. A compressor consists of a moving element enclosed in a housing. The unit is invariably coupled to a diesel-engine-driven or electric-powered prime-mover. Electric models are the most popular as they can be used indoors. The compressor, then, converts the mechanical power of its prime mover to pneumatic power using the compressible air medium. That is, it draws air from the surrounding atmosphere and pushes the air into a fixed-volume reservoir when driven by the prime-mover. The discharge pressure from the compressor must allow for pressure drops through downstream air treatment equipment, piping, and valves. Next, compressors are available in numerous types and have a wide range of power and pressure ratings.

However, it may be noted that the compression process in a pneumatic system is rather slow. Therefore, a sufficient quantity of compressed air must be invariably stored in the associated receiver tank.

Pressure Development in Compressors

Figure 4.2 shows a prime-mover-driven reciprocating compressor connected to a reservoir. The compressor consists of a movable piston enclosed in a cylinder. Assume that a compressor delivers three cubic-metre per minute (3 m³/min) of air to the reservoir having a volume of two cubic-metre (2 m³). Using Boyle's law, the pressure rise in the compressor can be calculated easily, and the values of absolute and gauge pressures against time are given in the associated Table. Remember, the development of pressure to the working level is not instantaneous, but it takes time.

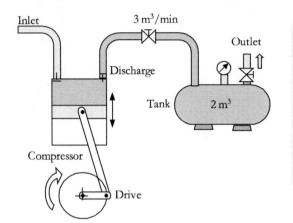

Time (Minute)	Nominal Volume (m³)	P, (ab) (bar)	P, (g) (bar)
0	2	1.0	0.0
1	5	2.5	1.5
2	8	4.0	3.0
3	11	5.5	4.5
4	14	7.0	6.0

Figure 4.2 | Pressure development in a compressor

Terms and Definitions, Compressor

A compressor is selected by the pressure it is required to operate at, the volume of air it is required to deliver, and the quality of air that is needed. The factors about a compressor, which are most important to pneumatic personnel, are its working pressure, operating pressure, theoretical and effective flow rates, drive unit, cooling methods, and regulation. These terms are explained below:

Working Pressure, Compressor

Working pressure is the pressure at the compressor outlet or in the receiver tank or in the associated pipelines. This pressure is usually higher than what is required at the operating position.

Operating Pressure, Compressor

Operating pressure is the pressure that is required at the operating position.

Specifications for Various Conditions of Air

The flow rate of compressed air in a pneumatic system is an essential parameter for sizing the components of the system. Remember, the reference conditions of air are different at different locations or under different situations.

The flow rate of a compressor is to be expressed under specified conditions of the air for its accurate representation. The flow rate is generally defined in terms of the atmospheric conditions at any specified location, or a set of reference conditions of pressure, temperature, and humidity. Accordingly, there are many ways of specifying the conditions of the air. The most important ways of representing the set of conditions of air for pneumatic systems are categorised under: (1) free air, (2) standard air, and (3) normal air.

Free Air

Free air is the air at the atmospheric conditions at any specified location, unaffected by the compressor. This term does not mean air under standard or normal conditions.

Air, under Standard Conditions

The standard set of reference conditions for representing the flow rate of air delivered by a compressor is specified in the ISO 1217 standard. This set of conditions is defined as a pressure of 1 bar(a), temperature of 20°C, and relative humidity of 0%.

Air, under Normal Conditions

The normal inlet conditions of a compressor are specified as a pressure of 1.01325 bar(a), temperature of 0°C, and Relative Humidity (RH) of 0%. The summary of reference conditions is given in Table 4.1.

Table 4.1 | Summary of reference conditions for air

Conditions of Air	Pressure	Temperature	Humidity
Free Air	Local conditions		
Standard Air (ISO 1217)	1 bar	20°C	0%
Normal Air	1.01325 bar	0°C	0%

The Flow Rate of Air, Compressor

The flow rate of air, in respect of a compressor, is the volume of air displaced or delivered per unit of time at the rated speed of the driveshaft. Note that the delivery is specified under the rated conditions of pressure, temperature, and relative humidity. Therefore, the flow rate can be measured in terms of: (1) Theoretical flow rate (or displacement) and (2) Effective flow rate (or delivery).

Theoretical Flow Rate (or Displacement Volume)

The theoretical flow rate is the quantity of inlet air that a compressor displaces. It is simply a mathematical calculation of the bore size, stroke, and drive speed of the compressor, irrespective of variables like temperature, atmospheric pressure, humidity, etc. The theoretical flow rate of a compressor is the product of the volume of air swept in one revolution of its drive shaft and the number of revolutions per unit of time. It is usually expressed as litres per minute (lpm) or cubic feet per minute (cfm). Also note: 1 cfm = 28.32 lpm or 1 lpm = 0.0353 cfm.

Effective Flow Rate (or Delivery Volume)

An effective flow rate is the quantity of air that a compressor delivers at the specified discharge pressure (typically specified at 6 bar) over a period. The quantity of the delivered compressed air is usually converted back to the actual inlet atmospheric conditions of the compressor at a given site or the standard (or normal) atmospheric conditions, to normalise the effective flow rate. Accordingly, the effective delivery volume can be expressed in terms of the actual delivery volume [Free Air Delivery (FAD)] or the standard (or normal) delivery volume.

Actual Delivery Volume (Free Air Delivery):

It is the volume of compressed air delivered by a compressor at the specified discharge pressure (typically 6 bar) over a period. It is usually stated in terms of the actual prevailing atmospheric conditions at the inlet of the compressor. In other words, it is the expanded volume of air that a compressor forces into the associated system per unit of time. It is expressed in terms of lpm (fad) or cfm (fad).

To calculate the free air delivery of a compressor, firstly the atmospheric pressure (P1), the actual temperature (T1), and the humidity (RH1) at the inlet of the compressor are measured. Next, the maximum working pressure (P2), discharge temperature (T2), and the volume of the compressed air (V2) at the outlet are also measured. Pv is the saturation vapour pressure of the moist air. Finally, the volume V2 is referred back to the inlet conditions using the equation of the ideal gas. The value of V1 is the free air delivery of the compressor. (See Figure 4.3)

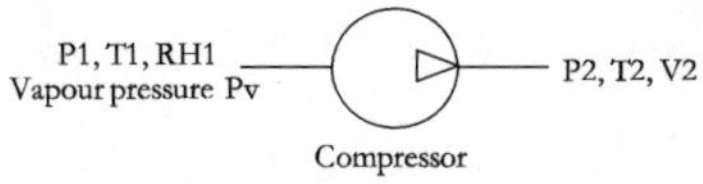

Figure 4.3 | Parameters to determine the free air delivery

$$\text{Free air delivery (FAD)}, V1 = \frac{P2 \times V2 \times T1}{[P1 - (Pv \times RH1)] \times T2}$$

Problem 4.1

How much air can be delivered by a receiver tank of 30 litres containing air at a pressure of 6 bar, under free air conditions? The temperature inside the tank is 40°C. Neglect the relative humidity.

Solution

$P1 = 6$ bar $= 7$ bar(a)
$V1 = 30$ litre
$T1 = 40°C = 273+40 = 313$ K
$P2 = 1$ bar(a)
$T2 = 20°C = 273+20 = 293$ K

$$V2 = V1 \cdot \frac{P1}{P2} \cdot \frac{T2}{T1}$$

$$V2 = 30 \cdot \frac{7}{1} \cdot \frac{303}{313} = 203 \text{ litres (Free air)}$$

Standard (or Normal) Delivery Volume

It is the volume of compressed air delivered by an air compressor at the specified discharge pressure and normally stated in terms of the standard (or normal) atmospheric air conditions. It is expressed in terms of lpm (std or normal) or cfm (std or normal).

Problem 4.2

How much air can be delivered by a receiver tank of 30 litres containing air at a pressure of 6 bar under standard air conditions? The temperature inside the tank is 40°C. Neglect the relative humidity.

Solution

$P1 = 6$ bar $= 7$ bar(a)
$V1 = 30$ litre
$T1 = 40°C = 273+40 = 313$ K
$P2 = 1$ bar(a)
$T2 = 20°C = 273+20 = 293$ K

$$V2 = 30 \cdot \frac{7}{1} \cdot \frac{293}{313} = 197 \text{ litres (Std)}$$

Problem 4.3

How much air can be delivered by a receiver tank of 30 litres containing air at a pressure of 6 bar under normal air conditions? The temperature inside the tank is 40°C. Neglect the relative humidity.

Solution

$P1 = 6$ bar $= 7$ bar(a)

$V1 = 30$ litre

$T1 = 40°C = 273+40 = 313$ K

$P2 = 1.01325$ bar(a)

$T2 = 0°C = 273$ K

$$V2 = 30 \cdot \frac{7}{1.01325} \cdot \frac{273}{313} = 181 \text{ Nlitres}$$

Kilowatt (kW) or Horse Power (hp) Rating, Drive

The kilowatt (kW) or horsepower (hp) rating of the drive motor or engine of a compressor is the work capacity of the drive. As a general rule, the higher the kilowatt or horsepower rating of a compressor drive, the greater the lpm or cfm output of the compressor for a given discharge pressure level.

Classification of Compressors

Compressors can be classified based on the design of their moving elements, the number of stages of the compression process, types of displacements, etc. Further, compressors can be of lubricated or non-lubricated (oil-free) designs. A brief explanation of various classification methods is presented below.

Classification: Reciprocating Vs Rotary Compressors

Compressors can be classified according to the specific design of the internal moving element used to create the flow of air. That is, they can be of the reciprocating type or rotary type. In a reciprocating compressor, the moving element like a piston reciprocates. Reciprocating piston compressors are very common and provide a wide range of pressures and delivery volumes. However, they are ideally suited for intermittent duty cycles.

In a rotary compressor, the moving element like, a set of screws or vanes, rotates. Rotary screw or rotary vane types are extensively used in many industrial applications. They are usually employed for continuous duty cycles. The type of compressor to be selected depends on the output requirements, specifically the duty cycle.

Classification: Single-acting Vs Double-acting Compressors

In the single-acting design of a piston-type compressor, compression of air takes place on one side of the piston, as shown in Figure 4.4. In this design, the compressor discharges one pulse of compressed air per revolution. Single-acting reciprocating piston compressors are commonly air-cooled and available in sizes up to 110 kW.

A double-acting piston compressor makes use of two sets of chambers and valves. It produces compressed air in two strokes per revolution of its crankshaft. This arrangement results in almost twice the capacity of a single-acting design of identical bore and stroke. Double-acting compressors are generally water-cooled and available typically in sizes up to 370 kW. They are expensive and entail high installation and maintenance costs.

Classification: Single-stage Vs Multi-stage Compressors

The term single-stage compressor indicates that an increase in pressure takes place in only one cylinder of the compressor. Single-stage compressors are generally used for pressures up to 12 bar. Single-stage compressors are adequate for small shops.

For higher pressures, multi-stage compressors are to be used. In a multi-stage compressor, the exhaust of one cylinder feeds the in-stroke of another to obtain higher outlet pressures. Each stage contributes some degree of compression. It is usually provided with an intercooler to remove the heat of compression. Cooling between stages reduces the input work requirement and the volume of air to be handled by the next stage. Double-stage compressors can be used for getting pressures up to 60 bar.

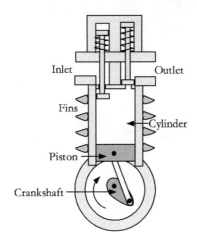

Figure 4.4 | A single-acting compressor

Classification: Oil-injected Vs Oil-free Compressors

An oil-injected compressor adds oil to the compression chambers for lubrication of its internal moving elements, sealing its internal clearances, and cooling its internal parts. However, some oil can get mixed into the air medium, and microscopic droplets of oil can carry through the machine and end up in the compressed air network, and eventually in the associated product.

If a process requires oil-free compressed air, then, an oil-free compressor is the best choice to ensure that there is no chance of oil contamination entering the compressed air network.

Size Classification, Compressors

Compressor sizes range from a small compressor generating less than 60 lpm with little or no preparation equipment to multiple compressor plant installations generating thousands of lpm. Compressors with delivery volumes up to 2400 lpm and drive powers up to 15 kW are considered small compressors. Compressors with delivery volumes between 2400 to 18000 lpm and drive powers between 15 to 100 kW are considered medium-sized compressors. Compressors above the medium limits are considered large compressors.

Classification: Positive Displacement Vs Dynamic Displacement Compressors

In general, compressors are classified, according to the compressing elements used, as: (1) positive displacement devices and (2) dynamic displacement devices. In a positive displacement compressor system, the air is confined within an enclosed space where it is compressed by decreasing its volume. In a dynamic displacement compressor, the air is accelerated by the rapidly rotating elements, such as rotor blades, causing some increase in pressure and a significant increase in velocity. A broad classification of compressors, according to the displacement, is shown in Figure 4.5.

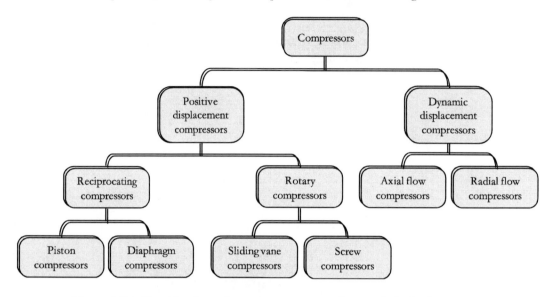

Figure 4.5 | Classification of compressors according to their displacements

Reciprocating Piston Compressor

Piston compressors are simple in design. Figure 4.6 shows the basic single-cylinder reciprocating compressor with the associated suction and delivery valves, inlet and outlet ports arranged on the head, and a drive with a crankshaft. The cylinder and head are typically made of cast iron for ruggedness and are provided with cooling fins for efficient heat transfer. A precision-machined alloy piston inside the cylinder moves back and forth when driven by the prime mover. The crankcase is usually made of high-grade cast iron.

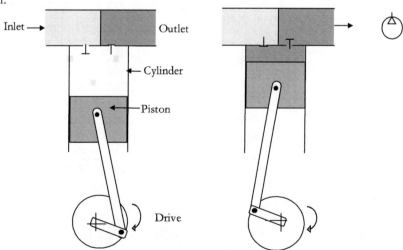

Figure 4.6 | Reciprocating compressor

As the piston moves during its inlet stroke, the inlet valve opens and draws air into the cylinder. During the outstroke of the piston, the air is compressed and discharged through the outlet valve. Suction unloaders, if provided, can run the compressor in unload mode when there is no demand.

Piston compressors are designed in single-acting or double-acting versions and for single-stage or double-stage operations. Multi-stage reciprocating compressors are more efficient and have a longer service life than single-stage compressors. Piston compressors are available as oil-free and oil-lubricated types. They can be provided with air-cooled or water-cooled heat exchangers as per the heat load. They must be designed for operator or plant safety as per the relevant standards.

They are employed where pressures in the range of 7 to 60 bar are needed. The free air delivery falls in the range of 60 to 1110 lpm. They are available with typical power ratings from 0.75 to 30 kW.

Piston compressors can provide an economical compressed air delivery to many industrial applications especially, in small-scale industries. They are most suited for intermittent duty cycles. Two-stage compressors with outsized storage are required for large industrial operations. High-pressure piston air compressors are widely used in the PET blowing, plastic, and packaging industries, military applications, power generating plants, and component test rigs.

Diaphragm Compressor

In piston compressors, there is a likelihood of introducing small amounts of lubricating oil from the piston walls into the compressed air. This minimal oil contamination may be undesirable in food, pharmaceutical, and chemical industries, as well as hospital and laboratory applications. For such applications, diaphragm compressors can be used as the power source.

A diaphragm compressor consists of a compression chamber with a flexible diaphragm and check valves, as shown in Figure 4.7. The diaphragm is linked to a crankshaft mechanism by a piston and a rod. The diaphragm separates the compressor chamber and the actuating piston. This feature allows the lubricating oil in the crankshaft to be excluded from the compressed air supply. The back-and-forth moving membrane, driven by a rod and a crankshaft mechanism, compresses air. However, diaphragm compressors have limited delivery and pressure levels. They are used most often for light-duty applications.

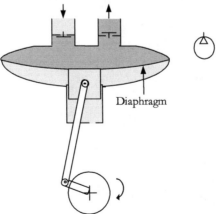

Figure 4.7 | Diaphragm compressor

Screw Compressor

Figure 4.8 shows a screw compressor. It consists of two helically grooved screws meshing with each other in an enclosure. Present-day screw compressors have asymmetric screw profiles for reducing internal leakage and improving energy efficiency. External gears are often used in screw compressors to synchronize the position of the screws with a negligible clearance of about 0.05 mm. No lubrication is required inside the compression chamber as the screws are not touching anywhere. Therefore, the compressed air delivered by screw compressors can be completely oil-free.

The design of the screws makes it possible to move air from the inlet to the outlet of the compressor. Compression is achieved by pushing the trapped air into a progressively smaller volume as the screws rotate. Since there are no surfaces that make contact with one another, this type of compressor does not necessitate cooling and is characterised by a low noise level and higher efficiency. They have the benefit of simplicity with fewer moving parts rotating at a constant speed. Another benefit is the steady delivery of compressed air without pressure fluctuations.

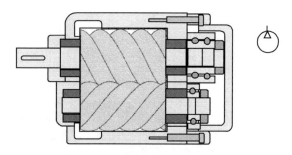

Figure 4.8 | Screw compressor

Screw compressors are optionally provided with integrated coolers and dryers. They are available in oil-flooded and oil-free technologies and with typical power ratings from 7.5 to 240 kW. Further, they can be coupled with fixed-speed drives or variable-speed drives. Oil-flooded screw compressors come with capacities up to 30000 lpm(fad) and oil-free screw compressors come with capacities up to 40000 lpm(fad). Oil-free models are designed for single-stage and two-stage operations. Single-stage screw compressors generally operate at pressures less than 10 bar. Higher pressures can be attained by multi-stage compression.

Rotary screw compressors are well-suited for continuous operation and are the lowest noise-producing compressors. A low sound level can be achieved by the optimized screw profile and vibration-isolating drive structure. Oil-free type rotary screw compressors can offer high-quality compressed air that may be required in some critical applications.

Vane Compressors

A vane compressor consists of a prime-mover-driven rotor with sliding vanes in close-fitting radial slots, as shown in Figure 4.9. The vanes are typically made of hard aluminium. The rotor moves within a larger circular cavity. The centres of the rotor and the cavity are offset by a certain distance, causing an eccentricity. The vane tips bear against the casing and form an adequate seal. Next, side plates are used to keep the air confined to the space existing along the width of the rotor and vanes. Oil is injected into the compression chamber to act as a lubricant as well as a seal. It also removes the heat of compression.

As the rotor rotates, the space between two successive vanes, at the suction side, increases. This expanding volume creates a partial vacuum, which draws air into the chambers formed by the vanes. The air is trapped in these chambers by rotation, and the trapped air is then moved through the compressor by the rotating vanes. As the space between the two rotating vanes decreases, compressed air is forced out through the discharge port.

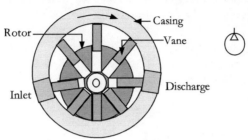

Figure 4.9 | Vane compressor

Rotary vane compressors are well-suited for continuous duty. Most rotary vane compressors are oil-lubricated. They produce low noise levels in the range of 72 to 80 dB(A). They are available in power ratings typically in the range of 5.5 to 630 kW, flow rate capacities from 1000 to 80000 lpm, and discharge pressures from 2.5 to 10 bar.

Rotary vane compressors are specially designed to meet the needs of the manufacturing, automotive, food and beverage, and pharmaceutical sectors.

Intake Filter, Compressor
An oversized intake filter eliminates coarse particulate matter from the ambient air at the suction side of a compressor system. This filter is usually a pleated paper type having large pores.

Cooling of Compressors
Cooling fins on smaller compressors permit the heat to be removed by radiation. A large compressor is usually equipped with an additional fan to take away the heat. In the case of a compressor plant with a drive power above 30 kW, forced air-cooling is not adequate. The compressors are then installed with a water-circulation cooling system.

Drive, Compressor
Electrical motors or IC engines drive compressors. The location of an application determines the type of drive. For example, generally, electric drives are ideal for indoor applications where permanent electricity is available, whereas, IC engines are preferred for outdoor applications. In factories, compressors are driven by three-phase or single-phase induction motors.

Power is transmitted from the motor or engine to the compressor through V-Belt, gear, or direct drive configurations. Traditional V-Belt drives provide great flexibility in coupling the compressor and its prime mover at minimum cost. A gear drive provides a reduction in the axial load on the moving element of a compressor, thus extending its operational life. Gear drives typically use less energy than V-belt drives. A direct drive can offer a compact configuration and minimum maintenance. Belts and coupling used in the compressor system must be adequately shielded for protection against accidents. Typical data for single-stage compressors and two-stage compressors are given in Appendix 2.

Storage of Compressed Air

As stated earlier, the slow response of a compressor to compress air to the required pressure calls for the storage of compressed air in a receiver tank. The use of air receivers is a simple method of power storage when the demand for air exceeds the capacity of the compressor. They assist in smoothening demand surges and help eliminate pulsation in the compressed air network. They also provide cooling of compressed air and consequent water collection. It may be noted that pressure in the receiver is generally higher than that required at the operating position.

Figure 4.10 shows the constructional features of an air receiver. A receiver tank is cylindrical to provide ample surface area. Air receivers are available as horizontal models and vertical models. Vertical models are used when there is less floor space for their installation.

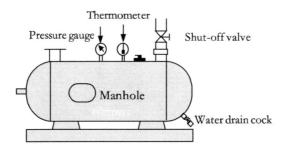

Figure 4.10 | Compressed air receiver tank

An air receiver is also provided with (1) a safety relief valve to guard against the development of high pressures arising from the failure of the pressure control scheme, (2) a pressure switch to sense the pressure inside the tank, (3) high-temperature switches for remote alarms, and (4) pressure gauges for pressure indication. A drain cock allows the removal of condensed water, and a cleanout opening allows entry for cleaning.

Air receiver tanks must be fabricated as per the relevant standard in one's region. Conformity to the standard ensures that the tank plates are of sufficient thickness, have proper materials, and are without defects. The tanks must be provided with openings and supports of the correct sizes. During the fabrication, proper welding techniques must be employed by experienced operators.

An air receiver can be installed downstream of the compressor to act as an air storage unit. Ideally, primary air preparation elements can be fitted upstream of the receiver to facilitate the filling of the receiver with partially cleaned and dried compressed air.

The advantages of compressed air receivers in pneumatic systems are enumerated below:
- Air receivers provide constant air pressure in a pneumatic system, regardless of fluctuating air consumption or compressor switching.
- The large surface area of the receiver tank cools the compressed air by way of dissipating heat quickly into the atmosphere. A fraction of the moisture in the air is then condensed and collected at the bottom of the receiver tank, which can be drained off subsequently.
- In the event of an electrical power failure, the air receiver can maintain the compressed air supply for some more time to carry on with the production.

Tank Sizes

The size of a compressed air receiver depends on the delivery volume of the compressor, load requirements, allowable pressure deviations in the receiver, etc. The larger the air receiver, the more compressed air is available for an application. The smaller the receiver, the more the compressor has to run to keep up with the demand. Typical tank sizes include: 160, 220, 420, 500 litres.

Sizing of Air Receivers

The size of the compressor tank can be determined based on the type of usage. If the usage is in short, quick, and concentrated bursts, then a small tank size can be used. If the unit is to sustain long periods of usage, a larger tank size should be selected. The size of a receiver tank (V) can be determined using the following formula:

$$\text{Receiver, size, } V = \frac{1.01 \times t \times (Q_r - Q_c)}{P_{max} - P_{min}}$$

Where,

V	= Size of the receiver tank, m^3
t	= Time to supply the required amount of air, min
Q_r	= Consumption rate of air, m^3/min (Std)
Q_c	= Compressor delivery rate, m^3/min (Std)
P_{max}	= Maximum pressure level in the receiver, bar
P_{min}	= Minimum pressure level in the receiver, bar

It is a common practice to increase the calculated size of the air receivers by about 25% for meeting unexpected overloads. It may further be increased by 1.5 to 3 times for satisfying the probable future expansion needs of the plant.

Problem 4.4

Calculate the minimum size of a receiver tank that must supply compressed air to a pneumatic system consuming 0.566 m^3/min (Std) for 6 minutes between pressures of 7 bar and 5.5 bar before the compressor resumes operation.

Solution

Q_r	= 0.566 m^3/min (Std)
Q_c	= 0
t	= 6 min
P_{max}	= 7 bar
P_{min}	= 5.5 bar

$$\text{Receiver, size, } V = \frac{1.01 \times t \times (Q_r - Q_c)}{P_{max} - P_{min}}$$

$$\text{Receiver, size, } V = \frac{1.01 \times 6 \times (0.566 - 0)}{7 - 5.5}$$

$$= 2.287 \text{ m}^3$$
$$= 2287 \text{ litres}$$

Problem 4.5

Calculate the required size of a receiver tank that must supply compressed air to a pneumatic system consuming 0.566 m³/min (Std) for 6 minutes between pressures of 7 bar and 5.5 bar, if the compressor is running and delivering air at 0.1415 m³/min (Std).

Solution

Q_r = 0.566 m³/min (Std)
Q_c = 0.1415 m³/min (Std)
t   = 6 min
P_{max} = 7 bar
P_{min} = 5.5 bar

$$\text{Receiver, size, } V = \frac{1.01 \times t \times (Q_r - Q_c)}{P_{max} - P_{min}}$$

$$\text{Receiver, size, } V = \frac{1.01 \times 6 \times (0.566 - 0.1415)}{7 - 5.5}$$

$$= 1.715 \text{ m}^3$$
$$= 1715 \text{ litres}$$

Compressor Controls

The operation of a compressor needs to be controlled to adjust the pressure and flow as per the demands of the application. Many types of control schemes are devised to suit many different requirements such as constant-speed or variable-speed or intermittent operation of the drive. The control scheme includes the load/unload method, inlet valve modulation, and motor start-stop control.

The load/unload method allows a compressor to run at full load or no load while its drive motor remains at a constant speed. The inlet valve modulation throttles the inlet of the compressor to change the delivery of the compressor to meet the varying demand for compressed air. The start/stop control starts or stops the drive motor of a compressor to store compressed air in the associated receiver tank and supply a sufficient quantity of compressed air to meet the application demand.

A constant-speed control can be employed for a compressor in a steady pneumatic system when the consumption of compressed air is more than 75% of the compressor delivery or the start-stop frequency of the associated motor exceeds the manufacturer's recommendations. The load/unload method and inlet valve modulation can be used for constant-speed control.

The start-stop (on-off) control can be used for a compressor when the compressed air consumption is less than 75% of the compressor delivery and an adequate compressed air storage facility is provided. This type of control can be used for applications where compressed air is not required continually, allowing the compressor to get sufficient cooling time. The start-stop control to regulate the pressure in a pneumatic system is explained in the following section.

A variable-speed drive can be employed in a screw compressor or vane compressor when energy efficiency is an important criterion. This type of control will adjust the speed of the drive dynamically to meet the varying demands of compressed air in an application.

A dual control system can be used for a compressor whose operation varies between intermittent and continuous duty according to the usage of compressed air. The switching between the constant-speed operation and the start-stop operation can be realised by a selector switch. Next, sequencing controls can be employed for a group of compressors to operate at peak efficiency levels.

Start/Stop Method of Pressure Regulation

Pressure development in a compressed air system needs to be regulated to prevent the over-pressurisation of the system. The schematic diagram of Figure 4.11 shows the components of compressed air generation and the most commonly used on-off pressure regulation. The compressed air generation part consists of a motor-driven compressor, a check valve, and a receiver tank. Remember, the check valve is used to contain the pressure in the receiver tank. The pressure regulation part consists of an electrical circuit with a contactor coil, a pressure switch, and on-off control elements.

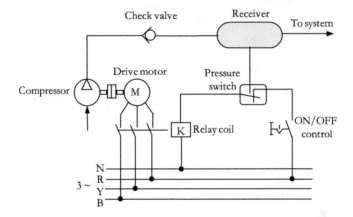

Figure 4.11 | The circuit diagram of an on-off pressure regulation method

The pressure switch senses the pressure in the tank. The upper and lower switching limits can be set on the pressure switch. That is, the switch opens when the upper limit of pressure is reached and it closes when the lower limit of pressure is reached.

Initially, the pressure is low in the receiver tank, and hence the pressure switch remains closed. When an ON signal is given, the control circuit is completed, and the contactor coil gets energised. Consequently, the three-phase motor is switched on and drives the compressor. The compressor, then, pushes air into the receiver tank and develops pressure.

When the pressure in the receiver tank reaches the upper limit, the pressure switch opens and interrupts the electrical control circuit. The motor, then, remains switched off and stops the pressure buildup.

However, the pressure in the tank reduces whenever the compressed air is consumed by the application. The pressure switch closes again after the lower limit of pressure is reached. The motor is, then, switched on again and drives the compressor.

This cycle of operation is repeated to regulate the pressure in the system.

To reduce the switching frequency of the motor, a large receiver tank and a large gap between the upper limit and the lower limit of pressure setting are necessary. Pressure variations in the tank depend on the permitted number of start/stop cycles per unit of time but normally lie within 1 bar.

It may be noted that the cost of producing compressed air at the standard 6 bar [90 psi] is considered to be the most economical. However, the cost would go higher for every rise in pressure.

Safety Relief Valve

A compressor is usually equipped with a relief valve to protect against any failure of its pressure regulation system. It is usually designed as a safety valve. The relief valve is normally closed and consists of a spring-biased poppet firmly pressed against the valve seat. Air pressure inside the compressor acts directly on the bottom of the poppet. When the air pressure is at an undesirably high level, the poppet can move off its seat, and air can exhaust through the valve.

Duty Cycle, Compressor

The duty cycle of an air compressor is the amount of time the compressor can run before it needs rest. For example, if the duty cycle of the compressor is 50% in one hour, then the compressor can run for a total of 30 minutes in any one hour.

Air Compressor Packaged Units

Most industrial air compressors are supplied as self-contained packages. An air compressor packaged unit is a fully assembled compact air compressor system, complete with an air compressor, air receiver tank, inlet filter, electric motor, belt/gear/direct drive, and automatic microprocessor controller. An electronic controller can be provided in an air compressor packaged unit for the intelligent shutdown of the unit and energy-saving operation. Optional elements include an aftercooler, particulate filter, dryer, automatic moisture drain, low oil safety control, magnetic starter, cooling fan, and pressure-reducing valve. The packaged units are available with low noise enclosure [65 – 70 dB(A)] and vibration isolators for their quiet operation.

Constructional Features, Compressors

Compressors are available for light-duty and heavy-duty applications. Some of the constructional features of compressors are highlighted in a most general way. They are typically constructed with cast-iron cylinders and hardened steel crankshafts with precision bearings. They are available as oil-free type or lubricated types. The drive part may be provided with an enclosed belt guard. Receiver tanks are constructed with steel tanks in vertical or horizontal designs. They have regulators, pressure switches, safety valves, fan-cooled motors, pressure gauges, and manual or automatic drains. A fully-packaged compressor includes an air-cooled aftercooler, a low-oil protection switch, and an electronic drain valve. The provision of vibration isolators can help reduce the operating noise and stress on the tank.

Applications, Compressors

From large industrial production machinery to smaller air tools like impact wrenches, spray guns, hammers, and even engraving pens, one air compressor can accommodate a range of different tools to service an entire operation. Some more applications of compressors are highlighted below.

High-pressure piston air compressors are widely used in the PET blowing, plastic, and packaging industries, military applications, power generating plants, and component test rigs.

Screw compressors are well-suited to heavy-duty industrial applications and high-power air tools when a large volume of high-pressure air is required continuously. They are widely used in automotive, metallurgy papermaking, food packaging, aerospace, mining, construction, and petrochemical sectors. Oil-free screw compressors are used in applications where entrained oil carry-over is not acceptable as in the semiconductor manufacturing industry.

Rotary vane compressors are specially designed to meet the needs of the manufacturing, automotive, food and beverage, and pharmaceutical sectors.

Comparison of Compressors

The process involved in the selection of a suitable compressor for a given pneumatic application depends on many factors. A generalisation is extremely difficult, but some significant features are identified by comparing the piston, screw, and vane compressors. Table 4.2 provides a comparison of various types of compressors in a most generalised way for educational purposes. For exact data, it is advised to refer to the manufacturer's catalogues.

Table 4.2 | Comparison of compressors

Criteria	Piston compressor	Screw compressor	Vane compressor
Moving elements	Reciprocating	Rotary	Rotary
Construction	Single or double acting, with deep-finned cast-iron cylinders	Two screws meshing with each other in an enclosure	A rotor with radial vanes eccentrically mounted in a stator
Lubrication	Oil-free / oil-lubricated	Oil-free/ oil-lubricated	Oil-lubricated
Compression stages	Single-stage/ two-stage	Single-stage/two-stage	Single-stage
Cooling	Cooling fan	Air- or water-cooled	Air- or water-cooled
Duty cycle	Mostly used for intermittent duty cycles	Continuous duty cycles	Continuous duty cycles
Pressure range	7 – 60 bar (or higher)	<10 bar	2.5 – 10 bar
Flow rate capacity	60 – 1110 lpm	Up to 40000 lpm	1000 – 80000 lpm
Power rating	0.75 - 15 kW	7.5 - 240 kW	5.5 - 630 kW
Drive speed	Fixed speed	Fixed-speed or variable-speed	Fixed-speed or variable-speed
Noise level	70 -75 dB(A)	45 – 66 dB(A)	72 – 80 dB(A)

Drain Traps

They collect and discharge liquids from after-coolers, separators, receivers, dryers, filters, and drip legs.

Installation of Compressors

A compressor must be installed taking into account various factors. A few guidelines for the proper installation of compressors are given below:

-Avoid dirty or dusty areas
-Avoid extreme temperature (hot or cold) conditions
-Ensure good ventilation
-Ensure adequate ambient airflow
-Provide good accessibility for maintenance and troubleshooting

Energy Cost, Compressed Air

The annual cost of electrical energy required to operate a compressor depends on the power rating of the compressor, the efficiency of the drive motor, the number of hours of operation of the compressor per year, and the electrical power rate per kWh (unit).

The relation for determining the electrical power cost is given below:

$$\text{Electrical Power cost} = \frac{(0.745 \times hp) \times (\text{hours per year}) \times \text{Power rate per kWh}}{\text{Motor efficiency}}$$

The annual cost of electrical energy can be one to three times the initial cost of the associated compressed air system, depending on the region and power cost. The energy cost of compressed air over an operating period is significant as compared to the capital cost of the compressor. Therefore, it is important to consider the energy efficiency when considering a new compressor.

Example 4.6 | Determination of Annual Energy Cost of Compressed Air

Calculate the annual energy cost of a 1-hp compressor used for an industrial production system running 8 hours per day (as in a single shift per day operation). The motor efficiency can be taken as 90% and the electrical power rate as $ per kWh (unit). Assume 300 days of operation of the system per year.

Solution

Compressor power rating	= 1 hp
Motor efficiency, η	= 90% = 0.9
Power rate per kWh	= $

Electrical power input	= Compressor power rating / Motor efficiency
	= 1/0.9 = 1.1 hp
	= 1.1 x 0.745 kW
	= 0.8195 kW

No. of daily hours of operation = 8
No. of annual days of operation = 300
No. of annual hours of operation = 8 x 300 = 2400

Annual energy consumption	= Electrical power input (kW) x Annual hours
	= 0.8195 x 2400 = 1966.8 kWh

Annual energy cost	= Annual energy consumption (kWh) x Power rate per kWh
	=1966.8 x $

Note: The reader may evaluate the annual energy cost of compressed air based on the local power cost.

The Selection of a Compressor

A compressor must be selected for meeting the requirements of the application. The selection of an air compressor for an application is based on the following important factors/parameters.

- The volume of free air per unit of time (lpm/cfm) required for all actuators and tools in the application must be determined. The flow rate requirement should be 10 to 20% more than the requirement of the application as calculated initially. The delivery of the compressor to be selected must meet the flow rate requirements of the application.

- The highest pressure required to operate pneumatic actuators and tools must be determined. The maximum pressure requirement decides whether the compressor to be selected should be a single-stage or a two-stage type. The pressure rating of the compressor may be taken 2 bar above the application requirement.

- The duty cycle of the intended application must be analysed. That is, whether the compressor is operated continuously or intermittently. For example, it is essential to use a rotary compressor or oversized piston compressor for an application with a continuous duty cycle. On the other hand, a machine, that requires a large burst of air for a small duration (intermittent duty cycle), can be provided with a piston compressor and a receiver tank large enough to supply the required amount of compressed air for the specified duration.

- The location (indoor/ outdoor, low altitude/ high altitude) where the compressor is to be located and the condition of atmospheric air (low temperature/ high temperature or low humidity/ high humidity) at that location also need to be considered while selecting the compressor. Remember, a compressor that is located at a high altitude or a high-temperature environment, or a high-humidity situation, gives less flow output than that specified by the manufacturer. Further, outdoor applications need protection against moisture and freezing.

- Electrical control requirements must be analysed and specified while selecting the compressor. Confirm the details such as voltage, phase, and frequency of the power source.

- The compressor to be selected must meet the noise level requirements of the application.

- The compressor to be selected must be simple to operate, energy-efficient, and easy to maintain. Note, the use of an inefficient compressor results in its reduced capacity utilisation and shortened service life, and higher energy costs.

Objective-type Questions

1. The volume of compressed air delivered by a compressor at the specified discharge pressure under the prevailing inlet condition is stated in terms of its:
   a) displacement volume
   b) free air delivery volume
   c) normal air delivery volume
   d) standard air delivery volume

2. Regulation of working pressure in a pneumatic system is made using:
   a) a pressure regulator with a vent hole
   b) a pressure relief valve
   c) a pressure safety valve
   d) on-off regulation or inlet valve modulation

3. Rotary compressors are well-suited for
   a) Higher presser levels
   b) Continuous operation
   c) Light-duty applications
   d) Intermittent duty cycles

4. Mark the <u>correct</u> answer
   a) A screw compressor can be a double-acting type
   b) A vane compressor can be used for obtaining minimal oil contamination
   c) A double-stage compressor produces compressed air on two strokes per revolution of its crankshaft.
   d) In a positive displacement compressor, the air is confined within an enclosed space where it can be compressed by decreasing its volume.

5. The compressor that excludes lubricating oil from the compressed air supply is:
   a) Vane type compressor
   b) Diaphragm compressor
   c) Double-stage compressor
   d) Double-acting type compressor

6. The compressor that is most suited for continuous operation and low noise levels is:
   a) Piston compressor
   b) Vane type compressor
   c) Screw type compressor
   d) Diaphragm compressor

7. The compressor that is not well suited for continuous operation and low noise levels is:
   a) Piston compressor
   b) Vane type compressor
   c) Single-stage screw compressor
   d) Double-stage screw compressor

8. Mark the <u>incorrect</u> answer:
   a) Rotary compressors with variable-speed drives can be used for energy efficiency.
   b) Power can be transmitted from the electric motor to the compressor through V-belt, gear, or direct coupling.
   c) A large rotary compressor is ideal for supplying a large burst of air for a small duration.
   d) An air receiver in a pneumatic system can reduce pressure fluctuations regardless of varying air consumption by actuating devices.

Review Questions

1. What are the primary functions of a pneumatic power source?
2. What function does a compressor serve in a pneumatic system?
3. What is the principle of operation of an air compressor?
4. Why is it required to prepare the compressed air delivered by a compressor?
5. Define the following: (1) Free air, (2) Standard air, and (3) Normal air.
6. Differentiate the working pressure and operating pressure in a pneumatic system.
7. Define the term 'displacement' and 'delivery' concerning an air compressor.
8. Explain the meaning of free air delivery (FAD).
9. What is the meaning of the effective flow rate of a compressor?
10. What are the ways to specify the effective flow rate of a compressor?
11. Define the term 'Two-stage compression' in respect of air compressors. What is the advantage of two-stage compressors?
12. Differentiate between rotary compressors and reciprocating compressors.
13. What are the two types of rotary compressors?
14. Differentiate between positive and dynamic displacement compressors.
15. Why is cooling required in reciprocating compressors?
16. Explain the purpose, constructional features, and applications of diaphragm compressors
17. When is a diaphragm compressor used?
18. Explain the constructional features of screw compressors.
19. What are the advantages of screw compressors?
20. Explain the constructional features and applications of diaphragm compressors
21. When is a vane compressor used?
22. What are the ways to remove heat from compressors?
23. Why is it necessary to store compressed air in pneumatic systems?
24. Describe the constructional features of an air receiver with a neat sketch.
25. Give a brief note on compressor drives
26. State three functions of an air receiver in a pneumatic system
27. Briefly explain the constructional features of air receivers.
28. An air reservoir of 3 cubic metres is connected to a compressor, which delivers 6 cubic metres of air per minute. What will be the pressure in the tank, measured by a pressure gauge immediately after three minutes? Ans: 6 bar(g)
29. What are the essential maintenance activities to be carried out in the case of an air receiver?
30. What are various control schemes available for compressors? Briefly explain
31. Give a brief note on the following: (a) Pneumatic power transmission, (b) Storage of pneumatic energy.
32. Explain the on-off regulation of the compressor with a neat sketch.
33. What are the factors upon which the annual cost of electrical energy required to operate a compressor does depend?
34. Highlight a few points on the selection of a compressor.

[Solution to problem 28 is given at the end of the book]

Answer key for objective-type questions:
Chapter 4: 1-b, 2-d, 3-b, 4-d, 5-b, 6-c, 7-a, 8-c

Chapter 5 | Contaminants and Primary Air Treatment

Contaminants can enter a pneumatic system through the air taken in by the system compressor. In industrial surroundings, air carries a large number of solid impurities, moisture, and oil particles. The contamination in compressed air also includes particles originating in the compressor, such as residues of lubricants and mechanically abraded particles. Contaminants are harmful to pneumatic systems. It is essential to have high-quality air to promote the reliable and efficient operation of pneumatic components.

Solid Contaminants
Solid impurities are industrial dust which includes iron, carbon, silicates, fibreglass, soot, and other abrasive materials.

Moisture
Moisture present in the air is in the form of water vapour that remains suspended in the given volume of air. It is difficult to remove moisture from the air if it is in vapour form.

Oil Particles
Oil is used as a lubricating or working medium in many industrial machines. Therefore, in industrial surroundings, air carries harmful oil particles that can be more than 10 mg/m^3.

Humidity
Humidity is the amount of moisture present in the atmosphere. It is usually expressed in terms of either absolute humidity or relative humidity (RH).

Absolute Humidity
Absolute humidity is the actual amount of moisture present in one cubic metre of air at a given temperature. For example, if 8.7 grams of moisture is present in one cubic metre of air at a particular temperature, say at 20°C, then its absolute humidity is 8.7 g/m^3 at 20°C. Absolute humidity is always temperature-dependent. [See Figure 5.1(a)].

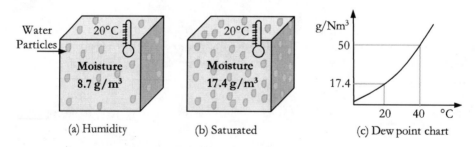

Figure 5.1 | Demonstrating the concepts of humidity

Saturation Quantity
A given volume of air at a specified temperature can contain moisture in the vapour form up to its saturation level. The saturation quantity of moisture is a function of temperature and is given by the dew point chart [See Figure 5.1(b)]. For example, the following can be observed from the dew point chart of Figure 5.1(c). At 20°C, one cubic meter of air can contain a maximum of 17.4 grams of

moisture. At 40°C, it can contain a maximum of 50 grams of moisture. It can be seen that the ability of a given volume of air to absorb moisture increases with an increase in temperature.

Relative Humidity

The relative humidity (RH) of air is the ratio of the absolute humidity of the air to the air saturation quantity at a given temperature. It is usually expressed as a percentage. That is,

$$RH = \frac{\text{Absolute humidity}}{\text{Saturation quantity}} \times 100\%$$

For example, if one cubic-metre of air contains 8.7 grams of moisture at 20°C, with a saturation quantity of 17.4 grams at 20°C, then the RH can be calculated as 8.7/17.4 x 100 = 50%.

It may be noted that 100% RH means the given volume of air is saturated. The relative humidity is dependent on both temperature and pressure. Decreasing the temperature or increasing the pressure will result in condensation of excess moisture above the saturation level.

Dew Point Temperature

Dew point temperature is the temperature at which dew begins to form as a result of condensation if the air is cooled at a constant pressure. It is the temperature at which the volume of air under consideration is saturated. The lower the dew point temperature, the more water will condense. Thus lowering the dew point temperature will reduce the moisture entrapped in the compressed air. For example, air with 17.4 grams of water vapour per cubic-metre has a dew point temperature of 20°C.

Air Quality Classification

ISO 8573-1: 2001 stipulates contaminants and quality classes of compressed air for general use. Air contains solid, water, and oil particles as contaminants. A quality class number is defined for each of these contaminants according to the permissible levels of parameters defining the contamination. These parameters and permissible values of them against each class are given in Table 5.1.

Table 5.1 | Permissible levels of contaminants as per ISO 8573-1: 2001

Class	Solids		Water	Oil
	Max. particle size (µm)	Max. concentration (mg/m³)	Max. pressure dew point (°C)	Concentration (mg/m³)
1	0.1	0.1	-70	0.01
2	1	1	-40	0.1
3	5	5	-20	1
4	15	8	+3	5
5	40	10	+7	25
6	-	-	+10	-
7	-	-	Not specified	-

A quality class 1.7.1 means that air contains solid particles of a maximum size of 0.1 µm and a maximum concentration of 0.1 mg/m³, an unspecified amount of moisture, and oil particles of a maximum concentration of 0.01 mg/m³.

Note: ISO 8573-1 2001 is superseded by ISO 8573-1 2010. The old version is presented here for easy learning purposes. The air quality classification ISO 8573-1 2010 is given in Appendix 5.

Preparation of Compressed Air

The compressed air delivered by a compressor has many harmful contaminants and objectionable conditions, as demonstrated in Figure 5.2. The contaminants and conditions are enumerated below:

- The compressed air is very hot.
- It contains a very high concentration of solid particles like dust, dirt, pollen, etc.
- It contains moisture in the vapour form as well as in the liquid form.
- It contains oil in the vapour form as well as in the liquid form.

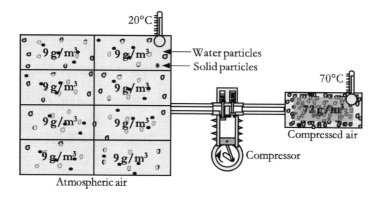

Figure 5.2 | Condition of compressed air at the outlet of a compressor

Effects of Contamination

Solid contaminants can damage compressor seals. They can disturb the operation of sophisticated downstream components such as valves and actuators. Water droplets resulting from the condensation can cause rusting of exposed surfaces, the formation of sticky emulsions, and consequent jamming of valves. It can also wash away lubricants from pneumatic components, resulting in faulty operation, corrosion, and excessive wear.

Stages of Preparation

To achieve any degree of reliability, the components of pneumatic systems must receive clean and dry air. Hence, air must be prepared or conditioned before it can be allowed to go into a pneumatic system. The compressed air is usually prepared in various stages, as shown in Figure 5.3.

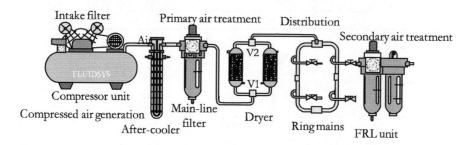

Figure 5.3 | Stages of preparation

The preparation of compressed air consists of reducing its temperature, removing solids, water and oil particles from it, regulating its pressure, and in many cases introducing lubricant in it.

Intake Filter
An intake filter is designed to protect the associated air compressor. It removes very large particles, usually of sizes greater than 200 microns, which can damage the compressor.

Primary Air Treatment
Primary air treatment is intended to reduce the temperature of the air at the outlet of the compressor, remove solid contaminants usually larger than 40 μ present in the air, and dry the air to reduce its humidity. The units used in the primary air treatment are an aftercooler, a main-line filter, and a dryer.

Aftercoolers
The air coming out of a compressor is very hot, and its temperature is typically between 80°C to 180°C. An aftercooler can be used to remove the excess heat and provide efficient cooling and thermal protection. It is intended to reduce the temperature of the compressed air to a certain value above the ambient temperature. Typically, aftercoolers are designed to achieve an approach temperature of 2.7°C, 5.5°C, 8.3°C, or 11°C above the cooling-medium temperature.

By reducing the temperature, most of the suspended water vapour in the air passing through the aftercooler will condense to a liquid state, and some oil vapour will fall out of suspension. The water can be trapped and drained off. Compressor manufacturers usually integrate aftercoolers within the compressor package. A standalone cooler is a separate unit installed downstream of the compressor. An aftercooler is sized according to the flow rate and the temperature of the air at the inlet of the aftercooler. Remember that the air is delivered by the compressor.

Types of Aftercoolers
There are two types of aftercoolers available in the market. They are:
- Air-cooled type
- Water-cooled type

Air-cooled Aftercoolers
Figure 5.4 shows an air-cooled aftercooler. It uses ambient air to cool the hot compressed air. The compressed air travels through finned tubes in the radiator while ambient air is forced over the cooler by a motor-driven blower. The forced air removes heat from the compressed air. As the air cools, the moisture condenses. This intermediate cooling helps to condense up to 75% of the moisture present in the compressed air into liquid water. The condensed water can be trapped in a moisture separator installed at the discharge of the aftercooler and can be drained off. Typically, air-cooled aftercoolers are sized to cool the compressed air to an approach temperature of 11°C above the ambient.

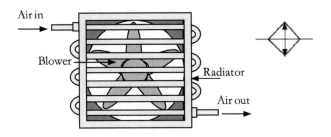

Figure 5.4 | Air-cooled aftercooler

Water-cooled Aftercooler

Figure 5.5 shows a water-cooled aftercooler. The standard style of a water-cooled aftercooler is the shell and tube type in which a bundle of copper tubes is fitted inside the shell. The hot compressed air flows through the tubes in one direction while the cooling water can preferably flow in the opposite direction around the tubes in the shell. Moisture in the compressed air turns into liquid water, as the compressed air cools. A moisture separator with a drain valve can remove the water.

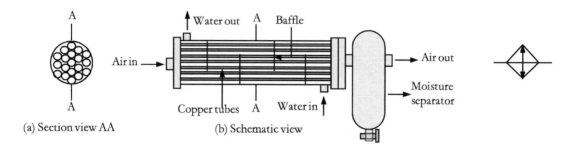

Figure 5.5 | Water-cooled Aftercooler

A modulating valve is recommended to maintain a consistent temperature and reduce the cooling water consumption. The tube bundles can be of the fixed type or removable type. Fixed tube bundles cost less but are more difficult to maintain as compared to removable bundles that can be detached for cleaning and servicing. The unit can be mounted in a vertical or horizontal configuration. The advantages of using a water-cooled aftercooler include: (1) better heat transfer and (2) there is no need for electricity. The disadvantages include: (1) high water usage and (2) difficult heat recovery.

Condensate Separator (Drain Trap), Aftercooler

As the compressed air cools, water condenses out, making the freshly produced compressed air very wet. A separator will collect and remove the liquids condensed by the aftercooler. When the air reaches the separator, centrifugal action causes the condensed water and other contaminants to hit the inner wall of the separator and then drip away to the drain. The separator is installed at the base of the aftercooler.

Condensate Separator with Automatic Drains, Aftercooler

Condensate can drip down the walls of the separator into the automatic condensate drain. When the drain fills with water, a float rises. As the float is lifted to a preset level, the valve opens to empty the condensate. This automatic action ensures that condensate and other contaminants do not build up in the cooler to pollute the compressed air supply.

Aftercooler Sizing Considerations

First, determine the compressor outlet temperature. The temperature at the outlet of a two-stage reciprocating piston compressor is typically about 120°C and that for a rotary screw compressor is about 90°C. Next, find the maximum free air delivery in terms of lpm or cfm and the pressure in terms of bar or psi.

After determining the parameters, a correctly-sized aftercooler can be selected from the size charts provided by manufacturers.

Compressed Air Filters

Filters can be fitted to the mainline and branch lines of a compressed air system to remove dust, dirt, water, and oil particles from the compressed air. They can be used to clean the air to a recognised compressed air purity standard, such as ISO 8573. The mainline filter must be sized to handle the maximum flow rate in the mainline and a branch line filter may be sized to handle only the flow rate in the branch line. Pre-filters and after-filters, most often used with dryers, must be able to handle the maximum flow rate through them. It is essential to change filters on schedule to reduce pressure drops.

Parts of Filters

A typical filter unit is shown in Figure 5.6. It consists of the following main parts: (1) Filter head, (2) Filter bowl, and (3) Filter element

Filter Head

A filter head holds the filter element and its housing. It consists of ports for the inlet and outlet. It may also consist of optional ports for the pressure gauge and visual and electrical indicators. It is made of zinc alloy or aluminium.

Filter Bowl

A filter bowl encloses the filter element. It confines the compressed air within the unit. It must be capable of withstanding the pressure within the unit. It is usually made of aluminium with a liquid level indicator or a transparent polycarbonate bowl with a metal guard.

Filter Element

The filter element is housed in the filter bowl. A filter element is usually made up of sintered plastic or sintered bronze. It consists of millions of tiny pores of micron sizes.

Additional Features of Filters

Pneumatic filters can be arranged with many additional features, such as manual/automatic drains and visual/electrical clogging indicators. Seals are usually made of Nitrile Buna Rubber (NBR).

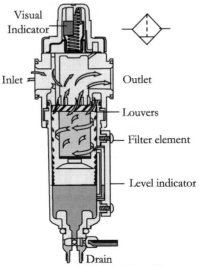

Figure 5.6 | Main-line filter

Working Principle of a Filter

Air entering the inlet port is directed to flow through angled louvres. This arrangement causes the air to spin as it enters the bowl. The centrifugal action of the rotating air causes large pieces of dirt and water droplets to be thrown against the inner wall of the filter bowl. These contaminants then flow down into the bottom of the filter bowl.

A baffle inhibits turbulent air from splashing water onto the filter element. The air, which has been pre-cleaned in this way, then passes through the filter element, where fine dirt particles are filtered out. The size of the dirt particles, which can be filtered out, depends on the mesh width of the filter cartridge. In coarse filters, mesh widths can be ≥40 microns. The compressed air exits through the outlet port.

Filter with Service Life Indicator

When a filter cartridge is clogged, its throughput is reduced and results in the wastage of energy. The extent of the pressure difference between the inlet and outlet will indicate the degree to which the filter element is clogged. Therefore, it must be cleaned or replaced, from time to time.

A filter is usually provided with a service life indicator, as shown in Figure 5.6, to monitor the pressure drop across the filter element and warn when the filter element needs to be replaced. The service life indicator may be a visual pop-up indicator or an electrical indicator. The visual indicator consists of a diaphragm, a green sleeve, and a movable red sleeve. With a new filter, the green sleeve is fully visible. The diaphragm is acted upon by the inlet and outlet pressures. As the outlet pressure drops and typically the pressure differential becomes 0.3 bar the diaphragm begins to move and lift the red sleeve covering the green sleeve. As the pressure differential increased further, the diaphragm moves up and the red sleeve progressively covers the green sleeve. When the pressure differential across the filter becomes 1 bar, the red sleeve covers the green sleeve fully. This position of the indicator is a warning to indicate that the set differential pressure has been exceeded. The filter element must then be replaced. The electrical service life indicator can be used for a remote visual and audible warning. It can also be used for turning off the associated machine or process automatically when the filter element is clogged beyond the acceptable limit. The use of a service life indicator allows for the replacement of the filter element only when its full dirt holding capacity is reached, rather than periodically replacing it on a time-scale basis.

Automatic Drain

The accumulated condensate at the base of the filter bowl should be expelled before the maximum condensate mark is reached; otherwise, it will re-enter the system. If a large amount of condensate accumulates in a short time, it is suggested to fit a zero-loss automatic drain in the filter in place of a manually operated drain cock to save energy. It uses a float to determine the level of condensate in the bowl. When the limit is reached, a control piston opens a valve seat that automatically ejects the accumulated condensate under air, through a drain line.

Types of Filters

Pneumatic filters can generally be classified as: (1) General purpose filters, (2) Coalescing filters, and (3) Adsorbing filters. General-purpose filters can remove solid particles down to 5 microns and water droplets from the system. Coalescing filters can remove particles down to 0.01 microns. It can provide air quality 1.7.2 as per the ISO standard 8573-1. However, it cannot remove oil vapours. The media in an adsorbing filter can attract and remove oil vapours. The adsorbing filter is an ultra-high efficient filter with an active carbon pack. The carbon pack assists in the removal of hydrocarbon gases. It can provide ISO air quality 1.7.1.

A coalescing filter is usually fitted ahead of a dryer. An adsorption filter is usually protected with an upstream coalescing filter. A particulate filter is recommended for the downstream of the dryer. With the combination of all three types of filters, it is possible to get high-purity compressed air.

Coalescing Filters

They are used for applications where the air is to be exceptionally clean and free of oil, as in food and drug processing, air bearings, and paint spraying. A coalescing filter element is shown in Figure 5.7. Air enters the inside of the filter element and passes through the filter media to the outer surface. The filter media is the borosilicate glass microfibre. The pathways through the media are so fine and intricate that the oil aerosol particles cannot pass through the media without contact. The particles coalesce (join together) when they contact the element media. The oil soaks and drains to the bottom of the sock where it drips into the bowl.

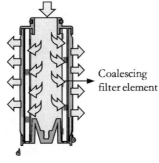

Coalescing filter element

Figure 5.7 | Coalescing filter element

Dryers

The natural water vapour content of air is concentrated and is carried through the compression process as a vapour in high temperatures. For simple applications, all that may be essential is an aftercooler, an air receiver, and a filter with condensate traps to remove the excess humidity. Additional means of dehydration must be provided where demands of high-quality compressed air are entailed. The most commonly used methods of compressed air drying are: (1) Absorption dryer, (2) Adsorption drying, (3) Membrane drying, and (4) Refrigeration drying.

Each of these methods has its specific characteristics and will produce optimum results only if correctly used. It is worthwhile to use an aftercooler before any dryer to reduce the amount of work to be carried out by the dryer. A dryer is ideally fitted downstream of the compressor. Pre-filters and after-filers are fitted to dryers for their optimum performance.

Absorption Dryer

Figure 5.8 shows an absorption dryer. In the absorption dryer, an active chemical, like phosphoric pentoxide or calcium chloride, is used as the drying agent. The drying agent chemically reacts with the moisture present in the compressed air that is passing through it and forms a water compound. The water compound is collected at the bottom of the container from where it can be drained off. Dry compressed air is delivered out of the dryer. It may be noted that the drying agent is used up by the moisture.

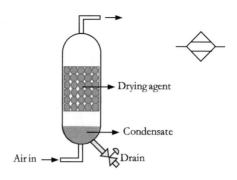

Figure 5.8 | Absorption dryer

Adsorption Dryer

Adsorption is the physical process of collecting moisture on the porous surface of certain granular materials such as silicon dioxide (Silica gel), activated alumina, copper sulphate, etc. Figure 5.9(a) shows the constructional features of a typical adsorption dryer. When compressed air is passed through a drying agent, like silica gel, the moisture present in the air is adsorbed by the drying agent. Dry compressed air is delivered out of the dryer. As more moisture is adsorbed, the drying agent gets saturated. Note that silica gel is a desiccant material as it has a high affinity for moisture. This method is used if an extremely dry air quality with pressure dew point down to -20°C, -40°C, or -70°C is required.

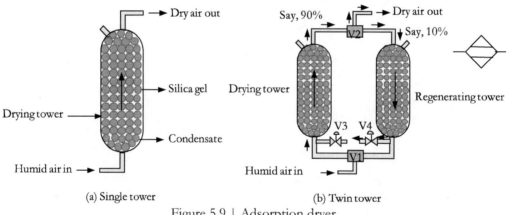

(a) Single tower   (b) Twin tower

Figure 5.9 | Adsorption dryer

The silica gel drying agent changes colour as it approaches its saturation point and reassumes its original colour when the moisture is driven out. However, the composition of the drying agent is not changed when it adsorbs moisture. Therefore, the moisture can be driven off by applying dry purge air or by application of heat, or both. Therefore, this method of drying is also known as regenerative air drying.

In practice, two parallel chambers are used, as shown in Figure 5.9(b), for non-stop production. While one chamber is drying the air, the other one can be set for regeneration after the pressure in the chamber has been reduced atmospheric pressure.

However, the capacity of the silica gel bed is limited due to the abrasion and contamination of the adsorption medium by oil and other substances. Under normal conditions, it is required to replace the drying agent once in 2 to 3 years.

When saturated, the drying agent can be renewed by blowing warm or cold air through the material, which then takes up the moisture. Accordingly, Adsorption dryers can be of the following three types: (1) Heatless type, (2) Heated type, and (3) Heated blower type.

Heatless Type Adsorption Dryer

In a heatless regenerative desiccant type dryer, as shown in Figure 5.9(b), no heaters are used. A heatless twin tower dryer diverts a portion of the dried compressed air to the off-line tower. This dry air then flows through the saturated desiccant and regenerates it. The purge air, now moisture-laden, is harmlessly exhausted through a muffler to the atmosphere. This type of dryer requires the lowest capital investment. However, this dryer technology may be more expensive to operate, because it requires a portion of the dried compressed air (up to 18%) to be diverted from the compressed air system for desiccant regeneration.

Heated Type Adsorption Dryer

In the heated type desiccant type dryer, dried air, diverted from the compressed air system, is first passed through a high-efficiency external heater before entering the off-line tower to regenerate the desiccant. Since heated compressed air can hold more moisture than unheated compressed air, only about 5% of the dried compressed air is needed for regeneration. Although the addition of the heater and associated components raise the initial capital investment for a heated dryer, less diverted compressed air means lower operating costs.

Heated Blower Type Adsorption Dryer

A heated blower-type dryer employs a high-performance centrifugal blower to direct ambient air through a heater and then through the off-line tower. The stream of heated air then regenerates the desiccant. The heated blower technology requires the highest initial capital investment. However, with no or little diversion of compressed air from the system for regeneration, it offers significantly lower operating costs than other types of desiccant dryer technologies.

Membrane Type Dryers

A membrane dryer consists of specially designed and bundled hollow membrane fibres. When compressed air passes through the membrane fibres, water vapour in the compressed air permeates the membrane walls. The dried air continues to flow to the downstream air system. The main drawback of this type of dryer is the relatively large amount of compressed air lost through the membrane walls along with the water vapour.

Refrigerated Dryer

The schematic of a typical refrigerated air dryer is shown in Figure 5.10. It consists of a heat exchanger and a refrigerating unit. In the first stage, the warm and humid compressed air is passed through the heat exchanger. The air gets cooled to a near-ambient temperature condition of the heat exchanger. The moisture present in the air gets condensed corresponding to the temperature in the heat exchanger and water is precipitated.

In the second stage, the partly prepared air is passed through the refrigerating unit to reduce the temperature of the compressed air to as low as 2°C. Again the moisture is condensed corresponding to the temperature in the refrigerating unit. The condensed water can be collected in the water traps provided at appropriate points. It may be noted that even oil particles get condensed when the temperature is reduced. The air is then filtered to remove the suspended solid particles and most of the oil mist contained in the compressed air. Finally, the air goes through the heat exchanger once again and gets discharged in a clean and dry condition.

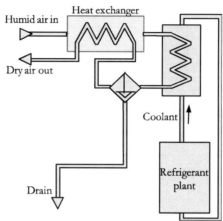

Figure 5.10 | Low-temperature dryer

Pressure Dew Point

The relevant operating pressure of the pneumatic system should be taken into account while comparing different methods of air-drying. The term 'pressure dew point' is used to make a distinction between the dew point at atmospheric conditions and that under operating conditions at higher pressures. This distinction is important because changing the air pressure changes the dew point temperature of the air. The pressure dew point is the lowest air temperature reached during the drying process at the specified operating pressure. For example, the typical pressure dew points that can be achieved with the adsorption dryers are −70°C, −40°C, and −20°C and that of the refrigerated dryers are +3°C, +7°C, and +10°C.

Selection of Dryers

The selection of dryers depends on variables, such as system demand, compressed air capacity, air quality requirements, and applicable life cycle costs that are unique to a compressed air system. . It may be noted that refrigeration dryers consume typically about 3% of the power that a compressor needs to produce the compressed air. However, desiccant dryers consume typically 10 to 25% of the power that a compressor needs to produce the compressed air.

Standards, Dryers

ISO 7183:2007 specifies the performance data that are necessary to state and applicable test methods for different types of compressed air dryers. It applies to compressed air dryers working with an effective (gauge) pressure of more than 0.5 bar, but less than or equal to 16 bar, and includes the data for the following types of dryers, membrane dryers, refrigeration dryers, or a combination of these.

A description of the principles of operation of dryers is given here within the scope of ISO 7183:2007. The standard identifies test methods for measuring dryer parameters that include the following: pressure dew point, flow rate, pressure drop, compressed-air loss, power consumption, and noise emission. The standard also provides partial-load tests for determining the performance of energy-saving devices or measures and describes the mounting, operating, and loading conditions of dryers for the measurement of noise. The standard does not apply to the following types of dryers or drying processes: absorption dryers, drying by over-compression, and integral dryers.

Comparison of Dryers
A comparison chart for compressed air dryers is given in Table 5.2

Table 5.2 | Comparison of dryers

Factor	Adsorption dryers	Refrigerant dryers
Process	Compressed air is passed through two identical drying towers each containing a desiccant bed (Silica gel) alternately. Warm and saturated moisture present in compressed air is adsorbed by desiccant material until it is saturated. While one tower is drying the compressed air, the other one is regenerating the saturated silica gel. Purge air is used to regenerate the desiccant.	Compressed air is cooled to a temperature as low as 3°C by a heat exchanger and a refrigerating unit. The moisture present condenses and can be separated and discharged from the system. The dried air is then reheated and discharged into the compressed air system.
Pressure dew points, typical	−70°C, −40°C, or −20°C	+3°C, +7°C, or +10°C
Advantage	Typically consumes 10 to 25% of the power that the compressor needs to produce compressed air.	Most economical type and typically consumes 3% of the power
Application	Used in an application that requires extremely dry air at quality class 1, 2, or 3	Used in an application that requires air quality class 4, 5, or 6

Objective-type Questions
1. The dew point chart is used to find:
   a) The maximum amount of moisture that can be absorbed by a given volume of air at a particular temperature
   b) the absolute humidity of a given volume of air
   c) the relative humidity of a given volume of air
   d) the pressure dew point of a dryer

2. Which component is used immediately after a compressor to reduce the temperature of the air discharged from the compressor:
   a) FRL unit
   b) Regulator
   c) Aftercooler
   d) Refrigerating unit

3. The primary function of an aftercooler in a pneumatic system is to:
   a) bring isothermal condition into the system
   b) pre-cool compressed air for the drying process
   c) cool compressed air to ambient condition before distribution
   d) reduce the temperature of the air delivered by a compressor to approximately 15 to 25°C above ambient

4. Identify the symbol:

Figure 5.11

a) Cooler
b) Dryer
c) Filter
d) Lubricator

5. A main-line filter in a pneumatic system
   a) removes only very large solid particles
   b) removes fine particles of sizes above 5 to 40 μ
   c) is capable of removing all particles including submicron particles
   d) is capable of removing coarse particles of mesh width greater than 100 μ

6. Mark the correct statement:
   a) Adsorption drying is a chemical process
   b) An absorption dryer uses the silica gel drying agent
   c) Pressure dew point is a parameter to compare various air-drying methods
   d) The low-temperature dryer is the most economical drying method under hot climatic condition

7. Identify the symbol:

Figure 5.12

a) Cooler
b) Dryer
c) Filter
d) Lubricator

8. A refrigerated compressed air dryer can yield a pressure dew point as low as:
   a) +20°F [-6.6°C]
   b) +35°F [+3°C]
   c) +50°F [+10°C]
   d) +70°F [+20°C]

9. A desiccant-type compressed air dryer can yield a pressure dew point as low as:
   a) -70°C [-94°F]
   b) -40°C [-40°F]
   c) -20°C [-4°F]
   d) -10°C [+14°F]

10. Desiccant dryers reduce dew point by:
    a) cooling compressed air to remove moisture by using a refrigeration cycle
    b) flowing wet compressed air through silica gel beads
    c) the chemical reaction of moisture in compressed air with drying agents
    d) passing compressed air through a bundle of hollow membrane fibers

11. ISO quality class 4.2.1, as per ISO 8573-1 (2010) is equivalent to:
    a) 1 to 5 μ particles < 10000, -94°F (-70°C) dew point, and <0.01 mg/m³ oil filtration
    b) 1 to 5 μ particles < 10000, -40°F (-40°C) dew point, and <0.01 mg/m³ oil filtration
    c) 1 to 5 μ particles < 10000, -4°F (-20°C) dew point, and <0.01 mg/m³ oil filtration
    d) 1 to 5 μ particles < 10000, +14°F (-10°C) dew point, and <0.01 mg/m³ oil filtration

Review Questions
1. State the effect of moisture content as air is cooled.
2. Name two ways to express the amount of moisture present in the atmospheric air.
3. What is meant by the term 'saturation quantity' concerning the moisture content in the air?
4. If the amount of moisture present in a given volume of air is at the rate of 9 grams per cubic metre at 20°C, what is the relative humidity of the given air? (Assume saturation quantity is 18 grams per m³ at 20°C).
5. What are the contaminants usually present in the atmospheric air?
6. What are the harmful effects of contaminants, if present in compressed air?
7. Explain the different stages of preparation of compressed air.
8. Briefly explain the air quality classification as per ISO 8573-1: 2010
9. What is the function of an aftercooler?
10. Explain the operational and constructional features of air-cooled aftercooler.
11. Explain the operational and constructional features of the water-cooled aftercooler.
12. Why is the filtration of dust particles in a pneumatic system carried out in a graded manner?
13. Briefly explain the primary function of a main-line filter in a pneumatic system.
14. How are oil particles present in compressed air removed?
15. Name the three main methods of drying compressed air. What are their essential differences?
16. Explain the working of an adsorption dryer with a neat sketch.
17. What is the reason for using two drying chambers in parallel, in the case of a silica gel adsorption dryer?
18. Explain the functional and constructional features of a heatless-type adsorption dryer with a neat sketch.
19. Explain the functional and constructional features of a heated-type adsorption dryer with a neat sketch.
20. Explain the functional and constructional features of a heated, blower-type adsorption dryer with a neat sketch.
21. Explain the functional and constructional features of a refrigerated dryer.
22. What are the typical pressure dew points that can be achieved with adsorption dryers?
23. What are the typical pressure dew points that can be achieved with refrigerant dryers?

Answer key for objective-type questions:
Chapter 5: 1-a, 2-c, 3-d, 4-a, 5-d, 6-c, 7-b, 8-b, 9-b, 10-b, 11-b

Chapter 6 | Compressed Air Distribution System

The objective of an air distribution system, as shown in Figure 6.1, is to act as a leak-proof carrier of compressed air and limit pressure drops within permissible limits. The air distribution system is made up of conductors and fittings, which interconnect various components of a pneumatic system. The distribution of compressed air should be planned and executed carefully by taking into account the following considerations: (1) correct sizing of pipes, (2) choice of pipe materials, (3) pipe layout, and (4) the total cost of the conductor system. A well-organised industrial pneumatic distribution system is designed with correctly sized pipes and components, ensuring a minimum number of elbows and bends so that pressure energy is not unnecessarily wasted.

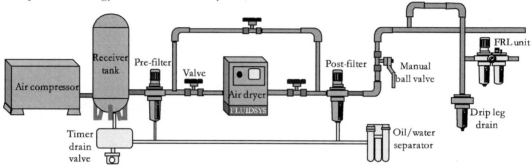

Figure 6.1 | A typical pneumatic air distribution system

Conductors
The conductors are generally divided into three classes: (1) Pipe (Rigid), (2) Tubing (semi-rigid or flexible), and Hose (Flexible). Many types of conductors may be used in the same installation.

Rigid Pipe
The main distribution system is made up of rigid pipelines, feeder lines, and associated fittings. Copper, iron, steel, and aluminium pipes must be brazed and welded or can be joined by way of threaded connectors to avoid the introduction of scale or welding particles into the system. Welded connections are robust and leak-free and are the main choice for fixed main distribution pipelines. As a rough rule, piping is employed for diameters above 50 mm.

Tubing
Tubing may be employed for conducting compressed air to air-powered tools and equipment, instruments, and gauges. Plastic tubing has gained wide acceptance in the industry for use as conductors in pneumatic systems as it is inexpensive and extremely easy to use with a high degree of flexibility. Food-grade tubes are colourless and tasteless and will not pass on extraneous flavour or odour to susceptible foods or beverages. There are both semi-rigid and flexible types of tubing.

Examples of semi-rigid types of tubing are aluminium, copper, and polyvinyl chloride, and flexible types of tubing are nylon and polyethylene. Nylon tubes are robust and can be used for a variety of applications. Polyurethane tubes are extra flexible and soft and are especially suitable in applications where short bending radii are indispensable. Since tubing can be bent, tubing assemblies require only a minimum number of fittings. A tube is usually specified by outside diameter and wall thickness. Nylon tubes are available in the following OD dimensions: 4, 5, 6, 8, 10, 12, 14, 16, 22, and 28 mm.

Hose

Hose assemblies are mainly used to connect a compressed air source to actuators that must be located on movable parts or because of the necessity to bend lines. They can be easily installed and require fewer installation skills than those for pipes or rigid tubing. They are capable of absorbing shocks and readily available in a whole range of pressure ratings.

Hoses are manufactured from synthetic rubbers and several plastics. They can be reinforced with fabric or wire braiding. A few examples of reinforcing hoses are polyester-reinforced PVC hoses and metal-braided rubber hoses. A hose is usually specified by its inside and outside diameters. It should have a smooth bore and must be resistant to oil vapours and lubricants. The walls of the hose must be sufficiently hard to resist heavy impacts and shock blows. The outer structure of the hose must be strong and abrasion-resistant.

Fittings

Pipes and tubes are joined to other pipes and tubes or the components of an installation by using some connectors. Push-in fittings are very compact units comprising retained collets and positive tube anchorage for easy tube insertion and hence for rapid assembly and are used for simple and quick assembly of pneumatic circuits. Figure 6.2 shows push-in and push-on fittings for pneumatic tubing.

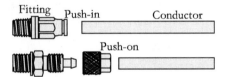

Figure 6.2 | Push-in and push-on fittings for tubing

Fittings are made of stainless steel, aluminium, bronze, or plastic with silicon-free nitrile rubber / Viton 'O' rings. They are available in a variety of shapes to form unions, elbows, tees, nipples, caps, plugs, couplings, crosses, etc.

Threads

Threaded pipe connections are available to a variety of standards, some of which are: American National Pipe Threads (NPT), Unified Pipe Threads (UNF), British Standard Pipe Threads (BSP), and Metric Pipe Threads (M).

The choice among these standards is determined by the standards already chosen for a user's region or country. Taper threads are cone-shaped. They form a seal between the male and female parts as they tighten, with assistance from some jointing compound or plastic tapes.

Threads on Fittings

The thread sizes on fittings and relevant standards are as follows:
- Taper male threads comply with ISO 7 (BS 21) and are designated by $R^1/_8$, $R^1/_4$, $R^3/_8$, $R^1/_2$, etc.
- Parallel male and female threads comply with ISO 228 (BS 2779) and are designated $G^1/_8$, $G^1/_4$, $G^3/_8$, $G^1/_2$, etc.
- Parallel metric threads are to ISO 261 (BS3643) designated M5x1, M10x1, M12x1.5, etc.

Threads on Ports

- The ports in components such as cylinders and valves are parallel and comply with ISO 228 (BS2779) these are designated $G^1/_8$, $G^1/_4$, $G^3/_8$, etc.
- Parallel metric threads are to ISO 261 (BS3643) designated M5x1, M10x1, M12x1.5, etc.

Quick-disconnect Couplings

Quick-disconnect couplings are widely used in pneumatic systems, mainly where there are frequent needs to uncouple the lines for maintenance, testing, and safety. Many disconnect couplings have double checks that can be used for easy detachment without any loss of compressed air.

Flow Resistance

The flow of compressed air through piping creates friction and consequent pressure drop. However, it should be acknowledged that differential pressure is essential for the flow of compressed air. The pressure loss is proportional to the square of the velocity of the flow.

Elbows, T-pieces, two-way valves, slide valves, etc., are also to be blamed for the interference with the flow and the corresponding loss of pressure. However, this pressure drop cannot be avoided but can be considerably reduced by routing pipes properly and assembling the fittings correctly.

Pipe Layouts

Various piping arrangements can be used in air distribution systems depending on usage requirements, the size of the plant, and delivery volume. The functions of an air-line system are to distribute leak-free compressed air, optimise flow, reduce turbulence, and maintain pressure. Generally, distribution is arranged as a manifold, as shown in Figure 6.3(a), or as a ring main, as shown in Figure 6.3(b), or cross-connected ring main, as shown in Figure 6.2(c).

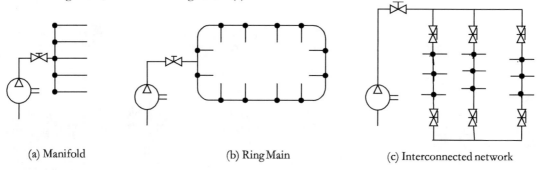

(a) Manifold   (b) Ring Main   (c) Interconnected network

Figure 6.3 | Pipe layouts for compressed air distribution

As the actuating devices consume air, the pressure is decreased at the downstream side of the line. One technique for reducing the pressure drop is to use the ring-main layout, by which any demand for compressed air can be met in two directions. A ring main ensures largely uniform pressure conditions in the air network.

With an interconnected network system, as shown in Figure 6.3(c), parts of the ring can be separated using the shut-off valves for maintenance, repair, and extension of the network without disturbing the rest of the system.

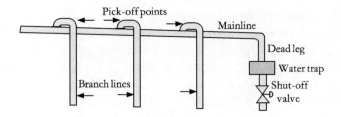

Figure 6.4 | Tapping of branch lines

The distribution pipe system can be considered as a part of the storage, and the compressed air inside is also subjected to external cooling. This cooling causes the moisture in the air to condense and consequently precipitate water. Hence, to provide drainage, the pipe should be inclined 1 to 2% downward in the direction of airflow, preferably to each corner.

All take-off points on the pipe for the branch lines, as shown in Figure 6.4, are tapped from the top of the pipe to prevent the entry of water into the branch lines. The condensate can be released from the system through a dead leg at the lowest point. An automatic drain valve can be provided for terminating the dead leg. Accumulated water can then be automatically drained off when pressure is on as well as when the system is shut down.

Heavy demands for compressed air are to be met occasionally at the ends of long lines, which can result in severe pressure loss. This pressure loss can be avoided by the installation of intermediate reservoirs as close as possible to the demand points. Thus, airpower can be stored close to where it is most needed.

Objective-type Questions

1. The purpose of the ring main air distribution system is to:
   a) ensure smooth flow of compressed air
   b) reduce pressure drop across the network
   c) ensure uniform pressure conditions in the air network
   d) remove water content in compressed air as far as possible

Review Questions

1. Show using a neat sketch the correct method of tapping an air main.
2. List four factors, which affect the flow of air through a pneumatic system.
3. What are the essential factors to be considered while planning and designing an air distribution system?
4. What causes the pressure drop when air is flowing through a pipe? How is it related to the flow velocity?
5. How can the pressure drop in the air distribution system be minimized?
6. Write a brief note on pipe materials.
7. Give a brief note on pneumatic tubing
8. What are the various methods of piping layouts used in the air supply system?
9. Explain the purpose of the ring main air distribution system and its advantages.
10. Mention the constructional features of the ring main air distribution system.

Answer key for objective-type questions: Chapter 6: 1-c

Chapter 7 | Secondary Air Treatment

Secondary air treatment is an effort to prepare compressed air finely, regulate pressure to the requirement of the application, and perhaps mix the air with a fine mist of lubricating oil, just before the entry of compressed air into the associated application. The components used in the secondary air treatment are Filter, Regulator, and Lubricator (FRL). The details of the filter are given in chapter 5.

Pressure Regulator

Pneumatic machines and appliances require relatively steady pressure for their satisfactory operation. However, the operating pressure tends to fluctuate due to variations in the supply pressure or load pressure. It is, therefore, essential to regulate the pressure to match the requirements of the load, regardless of the variations in the supply pressure or the load pressure.

A diaphragm regulator is a common type of pressure regulator found in an industrial pneumatic system. Figure 7.1 shows a pressure regulator. In general, a regulator consists of a body with an inlet, outlet, diaphragm, spring attached to the diaphragm, and valve stem and seat. The diaphragm has a centrepiece with or without a hole. Accordingly, the regulators are classified as relieving or venting type and non-relieving or non-venting type. In the relieving type of regulators, as shown in Figure 7.1, the centrepiece has a hole, and in the non-relieving type, the centrepiece is a solid part without a hole.

In the diaphragm regulator, a spring whose tension can be adjusted acts on one side of the diaphragm. The pressure is set by adjusting the spring tension. The operating pressure is indicated on a pressure gauge. The compressed air, which flows through a controlled cross-section at the valve seat, acts on the other side of the diaphragm. The diaphragm has a large surface area exposed to the downstream pressure and is quite sensitive to its fluctuations. The diaphragm moves as a reaction to the forces acting on it. Therefore, the diaphragm moves continually as the pressure fluctuates. The movement of the diaphragm controls the cross-sectional area at the valve seat and hence regulates the pressure.

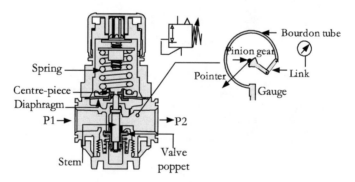

Figure 7.1 Pressure regulator with vent hole

Regulator with Vent Hole

When the load changes abruptly, the secondary pressure tends to become so high. This high pressure acting on the diaphragm can cause it to move a long way. This movement can close the valve seat entirely and can open the passage through the centrepiece, allowing the trapped compressed air on the secondary side to exhaust through the vent holes in the diaphragm and the regulator body.

Regulator without a Vent Hole

In a regulator without a vent hole, the compressed air cannot escape to the atmosphere, in the event of high back pressure acting on the diaphragm, as there is no exit path provided in the diaphragm for the trapped air.

Filter-regulator

In this design, the filter and regulator are combined as a single unit. Air flows first through the filter and is then directed to the regulator. The advantage of this design is that only one unit is to be mounted, thus simplifying the installation work and reducing the costs.

Pressure Gauge

Air Pressure is measured relative to one of the two references: (1) atmospheric pressure and (2) absolute vacuum. Atmospheric pressure is the most commonly used reference, as we operate in the atmospheric environment. Pressure measured relative to the atmospheric pressure is called gauge pressure.

The most commonly used devices for measuring pressure are bourdon tubes and diaphragm gauges. The well-known mechanical pressure gauge is the Bourdon tube gauge. (See Figure 7.1) A metallic tube of a non-circular cross-sectional area is formed into an arc. When pressure is applied to the inside of the tube, it forces the tube to expand. The higher the pressure, the greater will be the bending radius. This movement is transferred to a pointer through linkage, gear segment, and pinion, which indicates the pressure level on a circular scale calibrated in a proper unit of measurement.

Lubricator

The rapidly moving parts of valves and cylinders in a pneumatic system require lubrication to keep friction and wear to a minimum. Proper lubrication greatly increases the life of seals and wearing surfaces. The requirement of lubrication can be met by using valves and cylinders provided with a special lubricant or by injecting a controlled quantity of oil mist into the air stream using a mist lubricator. Mist lubricators can be of the micro-fog type or the oil-fog type. The micro-fog type is used for the most general-purpose pneumatic applications, and the oil-fog type is used for heavy-duty applications.

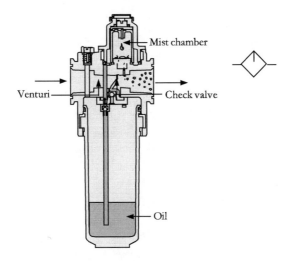

Figure 7.2 | Air-line lubricator

A mist lubricator is shown in Figure 7.2. It consists of a housing with an inlet port, an outlet port, a check valve, venturi (nozzle), an oil reservoir, a suction tube, an oil-regulating screw, and a mist chamber. The oil is usually stored in a transparent polycarbonate bowl or a metal tank of large volume with a sight glass.

Most lubricators operate on the Venturi principle. As compressed air is passed through the lubricator in the marked direction, the check valve opens, and the compressed air flows through the venturi. The restriction of the venturi creates a differential pressure between the bowl and the mist chamber, causing the oil to be drawn up through the suction tube to the point of low pressure. The oil then drips into the nozzle and gets atomised. As a result, a fine mist of oil is produced. The air stream then takes up the fine mist of oil. Heavy oil drops fall back into the bowl. The air mixed with the oil mist exits through the outlet port.

It should, however, be noted that excessive lubrication may produce a sluggish operation of valves and cylinders, malfunction of components, and increased pollution. A sight dome is provided at the mist chamber to indicate the delivery, and manufacturers specify the number of drops of oil per minute. Set it typically to 20 drips per minute at the flow rate of 10 litres per second. Metering of the oil is accomplished by a regulating screw, which provides a controlled orifice size.

The oil, when used up, can be filled up under normal pressure conditions through the hole provided for the purpose. Usually, a good quality, light-grade spindle oil will meet the lubrication requirement of most air systems. A lubricator cannot be filled under pressure except when fitted with a quick-fill device.

Air Service Unit

A combined FRL unit and its detailed and simplified symbols are shown in Figure 7.3. A handy and flexible method of combining these units is to use a modular system. The modular unit comprises the following:

- Shut off valves to isolate upstream air and downstream air
- Combined filter and pressure regulator with gauge
- Lubricator

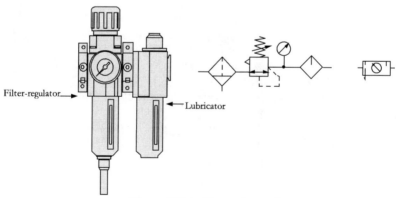

Figure 7.3 | Air service unit

These units are connected for the left-to-right flow with quick clamps, wall brackets, and pipe adaptors for rigidly fixing to the pipework. Individual units of the air service unit can be easily combined and slid into the pipework using quick clamps. These units can also be easily removed for servicing or replacement without disturbing the pipe joints.

Objective-type Questions

1. The mesh width of filters used in pneumatic systems for fine filtration is typically:
    a) below 200 microns
    b) in the range of 1 to 5 microns
    c) in the range of 5 to 40 microns
    d) in the range of 0.1 to 5 microns

2. Identify the symbol:

Figure 7.4

    a) Filter
    b) Lubricator
    c) Filter with manual drain
    d) Filter with automatic drain

3. Identify the symbol:

Figure 7.5

    a) Pressure regulator
    b) Pressure reducing valve
    c) Pressure sequence valve
    d) Unloading valve

4. The regulator in an FRL:
    a) reduces system pressure
    b) increases system pressure
    c) adjusts pressure to the requirement of individual machine
    d) unloads the compressor

5. Identify the symbol:

Figure 7.6

    a) Filter
    b) Regulator
    c) Lubricator
    d) FRL

6. Identify the symbol:

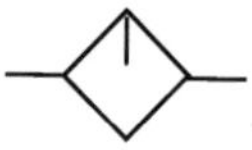

Figure 7.7

   a) Filter
   b) Regulator
   c) Lubricator
   d) FRL

7. The function of FRL is to:
   a) remove fine particles from compressed air
   b) set the pressure to the requirement of individual machine
   c) mix compressed air with an oil mist
   d) all of the above

8. The lubricator in a pneumatic system is:
   a) the first element in the power line
   b) the second element in the power line
   c) the third element in the power line
   d) the last element in the power line

Review Questions
1. What are the purposes of secondary air treatment in pneumatic systems?
2. Name the two contaminants removed from compressed air by filtration.
3. Explain the function of the filter component in the air service unit.
4. Explain the function of the regulator in the air service unit.
5. Differentiate between relieving type and non-relieving type regulators.
6. Explain the working of a bourdon tube pressure gauge.
7. What are the ways to realize the lubrication requirements of pneumatic systems?
8. Explain the function of the lubricator component in the air service unit.
9. What are the adverse effects of excess lubrication by a lubricator?
10. How excess lubrication by a mist lubricator be controlled?
11. What are the essential maintenance activities to be carried out for a filter?
12. Mention a vital maintenance activity to be carried out for a lubricator.
13. Why is it necessary to lubricate compressed air in many pneumatic systems?
14. Name four components that make up an air service unit.
15. Draw the symbols of (a) Filter, (b) Regulator, and (c) Lubricator.
16. Draw a simplified symbol for FRL to ISO 1219.
17. Give brief notes on the following:
    a) Coalescing filters
    b) Filter with service life indicators
    c) The Modular design of FRL

Answer key for objective-type questions:
Chapter 7: 1-c, 2-c, 3-a, 4-c, 5-d, 6-c, 7-d, 8-d

Chapter 8 | Pneumatic actuators

A pneumatic actuator is an output device that converts pneumatic power into mechanical power to drive the attached load to get some useful work. The resulting output motion can be either linear or rotary. Accordingly, there are two basic types of pneumatic actuators. They are: (1) Linear actuators and (2) Rotary actuators.

The linear actuators convert pneumatic power into straight-line mechanical power, and the rotary actuators convert pneumatic power into rotary mechanical power. An example of a linear actuator is a pneumatic cylinder and an example of a rotary actuator is a pneumatic (air) motor.

Linear Actuators
A linear pneumatic cylinder converts pneumatic power into a controllable linear force or motion or both. Technically and economically, pneumatic and hydraulic cylinders are the optimum form of linear actuators. It may be noted that a cylinder used in a pneumatic system provides a lower force output as compared to that provided by a cylinder of the comparable size used in a hydraulic system. However, a pneumatic cylinder, in comparison, can provide a higher speed output. Pneumatic actuators are available in a gamut of sizes, types, and mounting configurations to meet varied requirements of applications.

Working Principles of a Pneumatic Cylinder
Figure 8.1 shows the cross-sectional view of a typical pneumatic cylinder. It primarily consists of a barrel, piston, and piston-rod assembly, end-caps, and necessary seals and ports. The end-caps are securely fastened to the barrel. The piston with tight sealing forms two air chambers (piston chamber and piston-rod chamber) and can move within the barrel having two bearing surfaces. The cylinder is provided with two ports for permitting the flow of the compressed air. They are: (1) piston-side (cap-end) port X and (2) piston-rod-side (head-end) port Y.

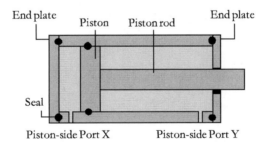

Figure 8.1 | A schematic diagram of a basic pneumatic cylinder

If compressed air enters the piston chamber through port X and the air in the piston-rod chamber is discharged through port Y, then the piston-and-rod assembly extends with a positive force (thrust). If compressed air enters the piston-rod chamber through port Y and the air in the piston chamber is discharged through port X, then the piston-and-rod assembly retracts with a definite force (pull). This type of pneumatic cylinder is called a 'double-acting' cylinder, as the extension and retraction of the cylinder are obtained pneumatically. If the motion of the cylinder is obtained pneumatically in only one direction, then such a cylinder is called a 'single-acting' cylinder. Note, the motion in the opposite direction can be realised by a spring or gravity or by any other external force.

Terms and Definitions
Some essential parameters concerned with the operation and applications of a pneumatic cylinder are its bore diameter, piston-rod diameter, force (thrust and pull), stroke length, speed, and piston-rod buckling.

Maximum Operating Pressure (P)
It is the pressure that overcomes all resistances in the system, which includes both useful work and losses. Alternatively, it is the maximum working pressure that the cylinder can sustain without adverse consequences.

Bore Diameter (D)
It refers to the diameter at the bore of the cylinder (See Figure 8.2). It can be used to calculate the bore area of the cylinder. It is also equal to the piston diameter, in a close-fitting hydraulic cylinder.

Piston-rod Diameter (d)
It refers to the diameter of the piston-rod of the cylinder.

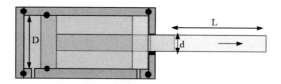

Figure 8.2 | Cylinder parameters

Stroke Length
The stroke of a cylinder is the distance through which its piston-rod moves. The volume of the cylinder to be occupied by air is related to the stroke length of the cylinder and is calculated by multiplying the stroke length by its piston area.

Maximum Stroke Length (L)
It is the maximum linear movement that a cylinder can produce. For standard models of double-acting cylinders, typically, the maximum stroke lengths can be up to 2000 mm, and for special designs, the stroke lengths can be up to 10 m.

Thrust
The theoretical thrust (out-stroke) or pull (in-stroke) of a cylinder is determined by multiplying the effective area of the piston by the working pressure. The effective area considered for the calculation of thrust is the full area of the cylinder bore and is given by $\pi D^2/4$. The effective area considered for the calculation of pull is the full area of the cylinder bore minus the rod area and is given by $\pi.(D^2-d^2)/4$.

$$\text{Thrust, F (Newton)} = \text{P (Pascal)} \times A_{ext} \text{ (m}^2\text{)}$$
$$\text{Pull, F (Newton)} = \text{P (Pascal)} \times A_{ret} \text{ (m}^2\text{)}$$

Where,

A_p is the piston area

A_r is the piston-rod area

A_{ext} is the active area during extension: ($A_{ext} = A_p$)

A_{ret} is the active area during retraction: ($A_{ret} = A_p - A_r$)

The present-day practice is to specify bore (D) and piston diameter (d) in millimetres and working pressure (P) in bar. In the formula for force, P is divided by 10 to express pressure in Newton per square millimetre [1 bar = (1/10) N/mm²]. The theoretical force (F) is given by:

$$\text{Thrust, F} = \frac{\Pi D^2}{4} * \frac{P}{10} \text{ Newton}$$

$$\text{Pull, F} = \frac{\Pi (D^2 - d^2)}{4} * \frac{P}{10} \text{ Newton}$$

Tables A3.1 and A3.2 in Appendix 3 give the theoretical forces of pneumatic cylinders in the SI system of units. Figures given in the tables do not make allowance for the loss due to seal and packing friction and air leakage. This type of friction is estimated to affect the thrust of the cylinders by about 10%.

Since the air pressure in a plant may vary erratically, due to intermittent use of large volumes of compressed air, the bore size of the cylinder must be large enough to provide the force required after allowing for any normal pressure drop. This sizing is critical as insufficient force may spoil the entire operation.

Example 8.1 | Determine the theoretical thrust and pull of a 50 mm bore double-acting pneumatic cylinder having a piston-rod diameter of 20 mm, supplied with compressed air at a pressure of 6 bar.

$$\text{Thrust, F} = \frac{\pi . 50^2}{40} . 6 = 1178 \text{ Newton}$$

$$\text{Pull, F} = \frac{\pi . \left(50^2 - 20^2\right)}{40} . 6 = 989 \text{ Newton}$$

Example 8.2 | Refer to the sample chart given in Table 8.1. Find the thrust of the cylinder with a 32 mm bore diameter at 7 bar.

Table 8.1 | Thrusts and pulls of double-acting cylinders

Cylinder bore mm	Piston rod dia mm	Thrust, N (at 6 bar)	Pull, N (at 6 bar)
25	10	294	246
32	12	482	414
40	16	753	633

Solution
The thrust of the 32 mm bore dia. cylinder at 6 bar [7 bar(a)] = 482 N

For the thrust for pressures other than 6 bar, multiply the thrust at 6 bar by the given absolute pressure and divide it by 7.

The thrust of the 32 mm bore dia. cylinder at 7 bar [8 bar(a)] = 482 x 8 /7 = 551 N

Cylinder Air Consumption

The equations for the volume of free air displaced by the piston during the extension stroke and retraction stroke of a double-acting cylinder are given below:

$$V(\text{out-stroke}) = \frac{\pi D^2}{4} \, S \, \frac{Ps + Pa}{Pa} \, 10^{-6}$$

$$V(\text{in-stroke}) = \frac{\pi (D^2 - \mathbf{d^2})}{4} \, S \, \frac{Ps + Pa}{Pa} \, 10^{-6}$$

Where,

D	=	Cylinder bore, mm
d	=	Rod diameter, mm
V	=	Volume of free air, dm³
S	=	Stroke, mm
P_s	=	Supply gauge pressure, bar
P_a	=	Atmospheric pressure (assumed to be 1 bar)
$\dfrac{(P_s + P_a)}{P_a}$	=	Compression ratio

The compression ratio $(P_s + P_a)/P_a$ may be considered as a multiplying factor to normalise the pressure condition. To estimate the total average air consumption of a typical pneumatic system, calculate the air consumption for each cylinder in the system using the formulae given above. Add the estimated air consumption of all cylinders and add 5% to make allowance for the loss due to leakage. Values of air consumption for forward and return strokes of pneumatic cylinders are given in Table A3.3 in Appendix 3.

Example 8.3 | Calculate the air consumption per mm stroke of a double-acting cylinder with 32 mm bore and 12 mm piston-rod diameter supplied compressed air at a pressure of 6 bar.

Solution
Bore diameter = 32 mm
Rod diameter = 12 mm

$$V(\text{out-stroke}) = \frac{\pi \cdot 32^2}{4} \cdot 1 \cdot \frac{6+1}{1} \, 10^{-6} = 0.00563 \text{ dm}^3/\text{mm}$$

$$V(\text{in-stroke}) = \frac{\pi \cdot (32^2 - 12^2)}{4} \cdot 1 \cdot \frac{6+1}{1} \, 10^{-6} = 0.00484 \text{ dm}^3/\text{mm}$$

Volume (Total) = 0.01047 dm³ per mm/cycle

Cylinder Speed

Assume that the piston-rod assembly of a cylinder moves with a velocity of 'v' when pushed by the system fluid with a flow rate 'Q'. Further, assume that the piston of area 'A' has moved a distance 'S' in time 't' for attaining the velocity v. Figure 8.3(a) and (b) shows two positions of the cylinder with the piston in position_1 and position_2 respectively for determining the cylinder speed during its forward stroke. Figure 8.3(b) also shows the position_1 superimposed over the position_2. Mathematically,

$$v = S/t \quad \text{or} \quad t = S/v$$

We can easily relate the theoretical flow rate (Q) of the system fluid to the speed (v) at which the rod moves if we consider the cylinder volume (V) that must be filled with the fluid and the distance (S) through which the piston must travel at the specified speed. We know that the volume (V) of the cylinder is the length of the stroke (S) multiplied by the piston area (A).

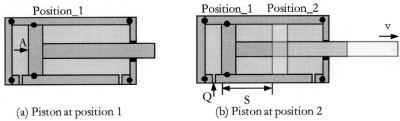

(a) Piston at position 1   (b) Piston at position 2

Figure 8.3 | Illustration of a cylinder in two piston positions

The flow rate (Q) to achieve the required speed (v), in the SI system units, is:

$$Q \ (m^3/s) = \frac{V \ (m^3)}{t \ (s)} = \frac{A \ (m^2) \ x \ S \ (m)}{t \ (s)} = A \ (m^2) \ x \ v \ (m/s)$$

It can be observed from the equation mentioned above that the speed (v) of a given cylinder depends on the flow rate (Q) of the system fluid. That is, a small-bore cylinder moves faster as compared to a large-bore cylinder with the flow rate remaining the same.

Cylinder Buckling

Very long-stroke cylinders are needed for some applications. If a compressive axial load is to be applied to the piston-rod of a cylinder, it must be within the safety limit to prevent rod buckling. Due to buckling stress, the permissible load of the rod that has a long stroke length is lesser than that ought to have been provided by the same maximum permissible working pressure and piston area.

Limitations on Maximum Thrust Force

The piston-rod diameter and overall length limit the maximum thrust force which a cylinder can practically provide. In cylinders with longer rods, the rod must be able to handle the thrust forces generated by the application. The cylinder must also be supported adequately.

Note that a head-end mounting provides greater column strength than the cap-end mounting, due to the smaller distance between the mounting points in the head-end mounting than that between the mounting points in the cap-end-mounting. The rod size of a pneumatic cylinder can be selected from the size charts, with the help of values of its free buckling length and the load imposed on the cylinder.

Example 8.4 | A pneumatic double-acting hydraulic press cylinder with an effective piston area of 19.625 cm² for push stroke, and a piston-rod area of 3.14 cm², operating at 6 bar does produce what theoretical forces for the push stroke and pull stroke?

Solution

Piston area, push stroke, A_{push}	= 19.625 cm²
Piston-rod area, A_{rod}	= 3.14 cm²
Pressure, P	=6 bar

Effective piston area, pull stroke, $A_{pull} = A_{push} - A_{rod}$
$$= (19.625 - 3.14) \text{ cm}^2$$
$$= 16.485 \text{ cm}^2$$

Thrust, F_{push}
$$= P \times A_{push}$$
$$= (6 \times 10^5) \times (19.625 \times 10^{-4}) \text{ N}$$
$$= 1178 \text{ N}$$

Pull, F_{pull}
$$= P \times A_{pull}$$
$$= (6 \times 10^5) \times (16.485 \times 10^{-4}) \text{ N}$$
$$= 989 \text{ N}$$

Example 8.5 | Refer to the sample chart given in Table 8.2. Find the total air consumption of the double-acting cylinder with a 100 mm bore diameter and 50 mm stroke length at 7 bar.

Table 8.2 | Air consumption of pneumatic cylinder

Bore mm	Rod mm	Air consumption for the		
		forward stroke of 1 mm at 6 bar dm³/mm	return stroke of 1 mm at 6 bar dm³/mm	combined strokes of 1 mm at 6 bar dm³/mm
100	25	0,05498	0.05154	0.10652
125	32	0.08590	0.08027	0.16617
160	40	0.14074	0.13195	0.27269

Note: 1 cubic decimetre (1 dm³) = 1 litre

Solution

Bore diameter	= 100 mm
Stroke length	= 50 mm

Air consumption for the combined strokes at 6 bar [7 bar(a)]   = 0.10652 dm³/mm

For air consumption for pressures other than 6 bar, multiply the air consumption value by the given absolute pressure and divide it by 7.

Air consumption for the combined strokes at 7 bar [8 bar(a)]   = 0.12174 dm³/mm

Air consumption for the combined strokes at 7 bar [8 bar(a)]   = 6.08685 dm³/cycle

Principal Parts and Constructional Details of Pneumatic Cylinders

Pneumatic cylinders are built in both single-acting and double-acting versions. They are also available in the 'sealed-for-life' type and 'serviceable' type. A cut-away view of a double-acting cylinder with cushioning devices is shown in Figure 8.4.

In the sealed-for-life type of cylinders, the piston may be pre-greased for life on assembly using a special grease and can be operated with non-lubricated air. The serviceable type of cylinders can be dismantled and reassembled by the user. The life of the cylinders can be extended by replacing worn seals. The factors concerning the design of a cylinder are its style, size, cost, duty, materials, mounting arrangements, and suitability for proximity sensing. The manufacturer's literature should always be checked to make sure that the cylinder is appropriate for the service conditions.

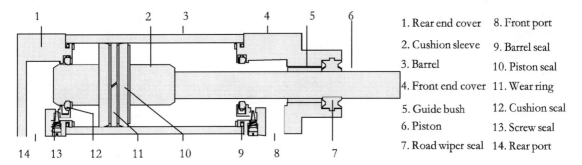

1. Rear end cover	8. Front port
2. Cushion sleeve	9. Barrel seal
3. Barrel	10. Piston seal
4. Front end cover	11. Wear ring
5. Guide bush	12. Cushion seal
6. Piston	13. Screw seal
7. Road wiper seal	14. Rear port

Figure 8.4 | Construction of pneumatic cylinder

Barrel

A cylinder barrel is generally made of a seamless drawn steel tube that is precision-machined to an accurate finish. The internal surface of the barrel needs to be very smooth to prevent wear and leakage. Adequate lubrication, for the cylinder, is essential. In such applications, where the cylinder is used occasionally or may come in contact with corrosive materials, aluminium, brass, or steel tube with a hard-chromed bearing surface can be used for the barrel. Unlike carbon steel, materials such as Brass and Aluminium are not subjected to corrosion. They are also better conductors of heat and assist in removing heat in high-frequency cyclic operations of the cylinder.

Piston

The piston in a cylinder is a loose fit inside the barrel and is held by a suitable seal/packing, providing a tight seal between high-pressure and low-pressure sides. It not only transmits force to the piston-rod but must also act as a sliding bearing in the barrel. Pistons are usually made of cast iron or steel.

Piston-rod

The surface of a piston-rod is exposed to the atmosphere when extended. Therefore, it is liable to suffer from the presence of dirt, moisture, and corrosion. While retracting, the settled particles on the rod may be drawn back into the barrel, causing harm to the precision parts inside.

Precision-ground stainless steel piston-rods are used in standard cylinders. Heat-treated chromium alloy can be used for the rod to reduce the effects of corrosion and for strength.

End Caps

Cylinder end caps are generally cast from iron or aluminium. For small-bore cylinders, glass-reinforced nylon may be used. They can be provided with corrosion-resistant materials. They also incorporate threaded entries for ports. The most important threads used in pneumatic cylinders are Metric thread, British Standard Pipe Thread, National Pipe Thread, and Unified Fine Thread. End caps have to withstand shock loads at the extremes of piston travel. These, end of travel shock loads can be reduced with cushion valves built into the end caps. The end caps can be fixed to the cylinder barrel by tie-rods, flanges, or threads.

Seals

Leakage from a pneumatic system can be a significant problem leading to wastage of energy, loss of efficiency, temperature rise, environmental damage, and safety hazards. Seals are used in cylinders to prevent the loss of energy in the form of leakage and to make effective use of the compressed air energy medium.

Characteristics, Seals

Seals are delicate and must be mounted with sufficient care. Dirt on a shaft or barrel can easily scratch a seal as it is slid into place. The essential characteristics needed for seals are: (1) long life, (2) low friction, (3) resistance to heat, (4) stability of form, (5) higher range of working pressure, (6) higher range of working temperature, and (7) good mechanical strength.

Classification, Seals

Based on the stress-conveying pattern, seals can be categorised into the following two types: (1) Static seals and (2) Dynamic seals.

Static Seals

Static seals are used to provide sealing between the stationary parts of a cylinder, for instance, between end caps and barrel. O-ring is probably the most commonly used static seal.

It is usually made of Buna as a standard material and Viton as an optional material. The moulded synthetic ring has a circular cross-section when unloaded. Application of the load causes the ring to be compressed at right angles, to give a positive seal (zero leakage) against two annular surfaces and one flat surface. O-rings are primarily used as static seals and may be used as piston seals in small-bore cylinders. The use of an O-ring as the piston seal in a medium or heavy-duty cylinder is not resorted to as any movement of the surface across the seal will cause it to rotate, allowing leakage to occur.

Dynamic Seals

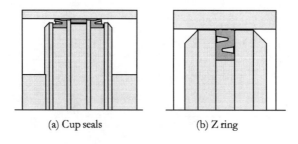

(a) Cup seals   (b) Z ring

Figure 8.5 | Dynamic seals

When a seal has to be provided between surfaces of which one of them is moving, a dynamic seal is most appropriate. Typical examples of the dynamic seal are the cup seal and the 'Z' seal. Pressure in the corresponding side of the cylinder makes the lip of the seal spread out and grip the barrel tightly to give a positive seal. The effectiveness of the seal increases with pressure, and leakage tends to be more of a problem at low pressures.

Cup Seal

A cup seal (cup packing) is shown in Figure 8.5(a). It is used on medium and large bore cylinders. They can provide sealing in one direction only. For single-acting cylinders, single-cup sealing is used, and for double-acting cylinders, double-cup sealing is used.

Z-Ring

A 'Z-ring' is shown in Figure 8.5(b). It is used on small-bore cylinders. It can provide sealing in both directions of airflow and has the advantage that it takes up only less space. The 'Z' shape acts as a light radial spring providing a light grip on the metal parts when un-pressurised. When air pressure is applied, the seal grip is automatically tightened.

Design of Seals

Seals are designed in different shapes according to different application requirements and are made of different materials to go well with varying environmental and operational conditions.

Piston Seals

An O-ring piston seal is used on small-bore cylinders. It is a loose fit in the groove with its outer diameter and the inner surface of the barrel. When pressure is applied, the 'O' ring is pushed sideways and outwards to seal the clearance between the piston and the barrel. For medium and large-bore double-acting cylinders double-cup seals can be used. They are cheap and easy to fit. However, they may easily be damaged by dirt.

Wear Ring

A wear ring is an open band fixed around the piston and is made from a hard plastic material or a good quality bearing bronze to provide the best wear resistance and excellent bearing support. In the event of high side loads acting on the piston, the wear ring becomes a bearing that prevents excessive distortion of the piston seals. It also guards the barrel against scoring by the piston.

Cushion Seal

Cushioning protects a cylinder and its load by absorbing the impact energy at the end of the stroke. Fixed cushioning with shock-absorbing pads can be fitted to small light-duty cylinders, which have, low mass in the piston, rod, and load. Large cylinders can be provided with adjustable air cushions. The basic idea here is to progressively slow down the piston as it approaches the end-of-stroke position.

Figures 8.6(a) and (b) illustrate the dual purpose of the cushion seal. It can perform the role of a seal when air flows in a particular direction and the role of a non-return valve when air flows in the opposite direction. When cushioning, the cushion seal restricts the exhaust air from the cylinder as the piston approaches the end-of-stroke position, thus performing the role of a seal, as shown in Figure 8.6(a). When air flows in the opposite direction, the flow is easily directed through the grooves in the seal, as shown in Figure 8.6(b).

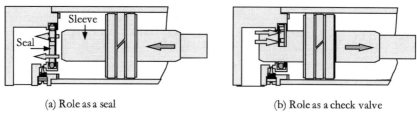

(a) Role as a seal	(b) Role as a check valve

Figure 8.6 | Cushion seal

Piston-rod Seal/Wiper Seal

A piston-rod seal is given in Figure 8.7(a). It is a special seal for harsh environments. It provides the dual role of pressure seal and wiper seal in a cylinder, thus enhancing the service life of the cylinder. A rod wiper is usually prepared from durable synthetic material. A classic example is the standard lip-type urethane wiper seal. The external body of the seal is a pressure-tight fit within the bearing housing. The cleaning action of the wiper seal removes the abrasive particles that can settle on the rod during the outstroke of the cylinder.

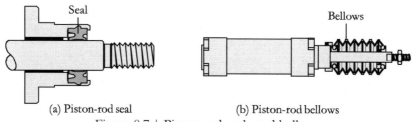

(a) Piston-rod seal	(b) Piston-rod bellows

Figure 8.7 | Piston-rod seals and bellows

Piston-rod Bellows

Figure 8.7(b) shows the piston-rod bellows. It is an alternative to the special wiper seals and is an ideal way to overcome the problem of the piston-rod getting scratched or abraded by falling debris when the piston-rod is extended.

Seal Materials

Seals are manufactured from a wide variety of materials. The choice of seals is determined by the following factors: (1) Type of fluid, (2) Operating pressure, and (3) temperature range.

Standard seals, made of synthetic rubber and plastic, are generally recommended for use in continuous running in the temperature range from +2°C to +80°C. Higher temperatures make the seals softer. So they wear quickly and produce added friction. Lower temperatures harden the seals and make them brittle. They are also liable to splitting and cracking. The most common present-day materials, along with their temperature range and characteristics, are listed in Table 8.3.

Table 8.3 | Temperature ranges and characteristics of seal materials

Material	Temperature range	Characteristic
Nitrile (Buna-N)	- 50°C to +100°C	Cheapest
Silicon	-100°C to +250°C	Expensive, tends to tear
Polyurethane	- 40°C to +200°C	Easy to use
Viton	- 20°C to +190°C	Most rigid
Teflon	- 80°C to +200°C	Most rigid

The Classification and Types of Pneumatic Actuators

Pneumatic cylinders as linear drive elements have been established on the operational level, very quickly and in many different ways. An important reason for this is the fact that there is no substitute for pneumatic actuators for core applications with a comparable cost-benefit ratio.

Most manufacturers bring new products continually to suit every industry sector from automotive manufacturing to onboard commercial vehicles, from rail applications to printing and textiles, from food packaging to the process industries, from the electronic sector to medical care, and in many other areas.

The availability of a wide range of cylinders provides users with the flexibility to meet their requirements precisely. A few of the essential types of actuators are categorised in Table 8.4.

Table 8.4 | Classification of pneumatic actuators

Main Type	Sub-type
Linear actuators	• Single-acting cylinder • Double-acting cylinder • Non-cushioned type • Fixed-cushioned type • Cushioned type
Special actuators	• Diaphragm cylinder • Magnetic type • Rodless type • Through piston-rod type • Non-rotating piston-rod type • Bellows actuator • Pneumatic muscle
Special assemblies	• Tandem cylinder • Multi-position cylinder • Impact cylinder • Telescopic cylinder • Hydro-pneumatic feed unit • Strip-feed unit • Rotary indexing table • Grippers (Finger-like, vacuum*)
Semi-rotary/ Rotary actuators	• Vane type (semi-rotary) • Rack-and-pinion type (semi-rotary) • Air motor
*Vacuum equipment	• Vacuum generator & suction cups

Table A3.4 in Appendix 3 gives the parameters for cylinders and rotary actuators.

Linear Actuators

Single-acting and double-acting cylinders are the two basic types of pneumatic linear actuators. They cover everything from 2.5 mm bore micro-cylinders to 320 mm bore standard cylinders.

Single-acting Cylinder

A single-acting cylinder is shown in Figure 8.8 with easily recognisable parts. It is a linear actuator with a barrel, piston, piston-rods, end caps, spring, seals, and only one port at the piston side for the passage of compressed air. An opening is provided at the rod end to breathe in and breathe out air to/from the spring chamber. The opening is usually provided with a gauze cover or filter to prevent the entry of foreign particles into the cylinder.

The cylinder can produce work only in one direction of motion when compressed air is applied, and hence it is named a single-acting cylinder. The return stroke of the piston is produced by the built-in spring when the compressed air supply is interrupted. The spring is designed to carry the piston and piston-rod assembly back to the normal position with adequately sufficient speed.

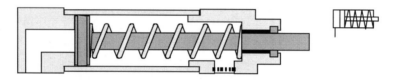

Figure 8.8 | Single-acting cylinder

In the mostly used sprung-in type design the spring is provided on the rod side of the single-acting cylinder. In the alternative design, known as the sprung-out type, the spring is provided on the piston side of the cylinder. The stroke length of the cylinder is limited typically to 100 mm due to the natural length of the spring. The exhaust air on the piston-rod side is vented to the atmosphere through the exhaust port.

The construction and simplicity of operation of the single-acting cylinders make them particularly suitable for applications such as clamping, cutting, ejecting, pressing, feeding, and lifting. For many applications the return forces required are minimal, thus enabling the spring return to be used. The use of single-acting cylinders minimises the air consumption and pipework/tube connection required. In the event of compressed air failure, a single-acting cylinder returns to its home position automatically.

Double-acting Cylinder

A double-acting cylinder is shown in Figure 8.9. It is a linear actuator with a barrel, piston, piston-rod, end caps, piston side port, piston-rod side port, and necessary seals.

The piston extends when compressed air is applied through the piston-side port provided that the chamber in the piston-rod side is exhausted. The piston retracts when compressed air is applied through the piston-rod-side port provided that the chamber in the piston side is exhausted.

The cylinder can produce work in both directions of motion when compressed air is applied and hence the name double-acting cylinder.

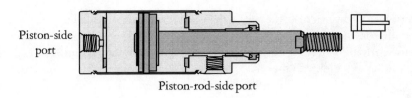

Piston-rod-side port

Figure 8.9 | Double-acting cylinder

For standard cylinders, bore sizes range from 1 mm to 320 mm. The force transferred by the piston-rod is higher for the forward stroke than that for the return stroke. Ideally, a conventional double-acting cylinder can be designed with unlimited stroke length. Still, practically the maximum stroke length is limited to about 2000 mm, due to the buckling and bending of the very long extended piston-rod.

Fixed Cushion Cylinder

Small-bore light-duty cylinders have fixed elastomeric cushions. The double-acting cylinder, as shown in Figure 8.9, has fixed cushions. These cylinders make use of synthetic rubber buffers to give a simple fixed cushion effect. These shock-absorbent discs, placed into the end covers, cushion the impact of the piston.

Air Cushion Cylinder

Cushions can be built into one end or both ends of a cylinder and can be of the adjustable type or non-adjustable type. Air cushion slows down a cylinder's piston movement just before reaching the end of its stroke.

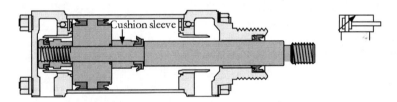

Cushion sleeve

Figure 8.10 | Adjustable cushion cylinder

Figure 8.10 shows a double acting cylinder with adjustable cushions. The cushion part consists of a throttle valve and a sleeve. Further, a cushion sleeve is attached to the piston. As the piston approaches the end of the travel position, the cushion sleeve blocks the usual exit for the air. This block forces the air to pass through the throttle valve which restricts the flow, progressively retarding the piston movement. The air cushions are usually designed to function over the final 2 cm of the stroke. The cushioning can be of the fixed type or can be adjusted by controlling the throttle valve. In the reverse direction, flow bypasses the throttle valve using a check valve within the cushion seal. With huge forces and high accelerations extra measures, such as providing external shock absorbers, must be taken to assist the load deceleration. Cushions are usually applied to cylinders whose piston speeds exceed 0.1 m/s.

Cylinder for Proximity Sensing

Figure 8.11 shows a cylinder suitable for proximity sensing. In this type of cylinder, a permanent magnet or magnetic band is provided around the circumference of the piston. The magnetic field is used to trigger the associated proximity (reed) switches/valves. These proximity switches can be fitted in the sensor groove or fastened to an integral rail on the cylinder. A magnetic cylinder is used in conjunction with magnetically-operated switches/valves. The switches/valves provide electrical/pneumatic feedback signals when influenced by the magnetic piston at the designated positions of the piston. The black band on the symbol indicates that the cylinder is suitable for proximity sensing.

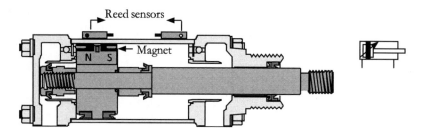

Figure 8.11 | Cylinder with magnet piston

Rodless Cylinders

A rodless cylinder consists of a barrel, piston, and carriage. The piston in the rodless cylinder can move freely in the barrel, as in conventional cylinders, when compressed air is applied. As there is no piston-rod in the rodless cylinder, there is no positive external connection or extension. Hence, for the power transmission, the piston is to be coupled to the carriage either magnetically or mechanically for loading.

Magnetic Coupling

In the magnetically coupled version, the piston and the carriage are fitted with sets of annular permanent magnets of opposite poles. Magnetic coupling occurs between the internal piston and external carriage. As a result, the carriage moves synchronously with the piston.

Magnetic coupling should be stronger by a certain safety factor as compared to the pneumatic thrust derived from the permissible working pressure and the piston area. This type of rodless cylinder is especially adaptable to an extremely long stroke length of up to 30 ft, because of the absence of a piston-rod. Another advantage of this design is that the cylinder barrel can be hermetically sealed from the outer carriage since there is no mechanical connection. Further, there is no possibility of leakage losses.

Mechanical Coupling

A mechanically-coupled rodless cylinder is shown in Figure 8.12. A full-length slot in the barrel joins the piston and external carriage in the rodless cylinder with mechanical coupling. Strips are continuously parted at one end and closed at the other end, as the piston moves through the stroke. Long sealing strips on the inside and outside of the cylinder tube prevent loss of air and ingress of dust. Adjustable cushions can also be provided to prevent end-of-stroke shocks.

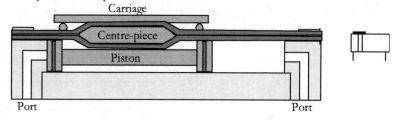

Figure 8.12 | A mechanically coupled rodless cylinder

Advantages of Rodless Cylinders
Rodless cylinders have many remarkable advantages over cylinders of conventional design. They are very compact, highly reliable, and occupy much less space in comparison to cylinders of other designs. There is no projecting piston-rod in a rodless cylinder. As there is no piston-rod in a rodless cylinder, there is no risk of piston-rod buckling. Another advantage of rodless cylinders is that extremely large stroke length is possible with this design. Further, the thrusts are the same in both directions of motion. For the accurate positioning of the carriage, the circuit for the rodless cylinder uses check valves to prevent the carriage from creeping.

Applications of Rodless Cylinders
Rodless cylinders are ideally suited for use in simple multi-axis handling devices. They also save a great deal of space with in-feed, removal, and transfer functions. There is no need to protect the piston against torsion.

Cylinder with Through Piston-rod
The cylinder has a piston-rod on both sides, as shown in Figure 8.13. The piston-rod is supported on two bearings. Therefore, the piston-rods provide better alignment and move smoothly through the openings. Further, the cylinder has equal areas on both sides of the piston and can give equal forces and speeds in both directions of cylinder motions.

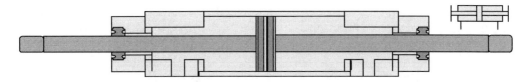

Figure 8.13 | Cylinder with through piston-rod

This type of cylinder is usually centre-mounted and can be used when the same type of task is to be performed at either end in a staggered manner. It can also be used when one end of the rod is to be used for transmitting force while the other end is to be used for tripping limit switches, control valves, or some other mechanism.

The through piston-rod can be provided with a hollow passage to conduct compressed air. A vacuum connection can be made to one end of this passage with the other end fitted with a suction cup.

Bellows Type Air Spring
Bellows-type air springs are single-acting actuators. They have one, two, or three convolutions in their flexible part. They also consist of Griddle rings and bead plates. Figure 8.14 shows a bellows-type air spring. The flexible part is a rubber-fabric component with an inner liner, reinforcement layers, and an outer cover. The flexible inner liner is in contact with compressed air and provides resistance to air permeability. The reinforcement layer is a structure of cords built into the flexible member to control its shape and strengthen its wall structure against internal air pressure. The outer cover protects the reinforcing cords from abrasion and weathering.

The girdle ring is a rubber-covered bundle of wires that restricts the diameter of the flexible member at the attachment point to form double or triple convolutions. Upper and lower bead plates are provided as end enclosures of the air spring to create an air-tight assembly.

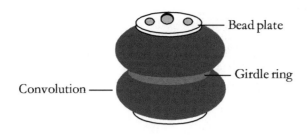

Figure 8.14 | Bellows-type air spring

They are generally made of thermoplastic or composite material, or corrosion-resistant aluminium, zinc, or steel, and supplied with blind taps and protruding bolts to facilitate their mounting to equipment.

Bellows actuators have no reciprocating parts or sliding seals, and hence they are virtually frictionless and maintenance-free units.

Standard models are manufactured with nominal diameters in the range from 50 mm to 950 mm with 1, 2, or 3 convolutions and stroke lengths up to 450 mm. Typically, the load capacity of air springs can vary from 400 to 400,000 N.

Working: When compressed air is applied to a bellows, it inflates and provides powerful short strokes. These movements can be used to lift loads. In the de-energised state, the bellows actuator is deflated. External restraints limit the maximum extension and compression of the bellows.

Applications of Air Springs: Air springs are very effective as single-acting cylinders, where short strokes and high forces are required, such as in clamping or pressing operations or where heavy loads are to be moved for short distances. They are often used in pneumatic spring systems and are ideal for absorbing vibrations of supported loads. For example, they provide solutions for the vibration-oriented mounting requirements of industrial machines as well as the suspension needs of commercial vehicles.

Pneumatic Muscle
Pneumatic muscle is a type of tensile actuator that mimics natural muscular movement. Tubing, which is impervious to fluids, is provided with a covering made of strong fibres in rhomboidal form. This structure forms a 3-dimensional grid structure, thereby reinforcing the tubing.

Figure 8.15 illustrates the operation of a pneumatic muscle. When air is allowed in, the grid structure changes in shape through radial expansion, and a tensile force is developed in the axial direction. The higher the internal pressure, the more the muscle is shortened. The maximum tensile force of a pneumatic muscle depends on its internal diameter.

The materials used for the reinforced tubing are chloroprene and aramide. Chloroprene is a plastic material of the elastomer group. Aramide is a reinforcing fibre made of aromatic polyamides.

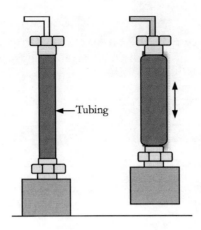

Figure 8.15 | Pneumatic muscle

Pneumatic muscles are available for the following nominal diameters: 10, 20, and 40 mm and the corresponding maximum stroke forces at 6 bar are: 400, 1200, and 4000 N respectively. A pneumatic muscle can be configured for nominal lengths up to 9000 mm.

Advantages and Disadvantages, Pneumatic Muscle
A pneumatic muscle can provide up to 10 times more initial force as compared to a piston-type cylinder with an identical diameter. It shows no stick-slip effect, as there is no piston inside the muscle. In a load-lifting system using pneumatic muscle, intermediate positions can easily be set by regulating the pressure.

The significant disadvantages of a lifting system with a pneumatic muscle are its inability to guide the associated load and overload protection. Another disadvantage of the pneumatic muscle is the ageing of its rubber material.

Applications, Pneumatic Muscle
The considerable force developed by a pneumatic muscle with its stick-slip-free movements makes it an interesting choice for many lifting tasks and in applications where accurate positioning at reduced speeds is required. Pneumatic muscle is a compact and powerful tensile actuator. It is ideally suited for gripping works.

Pneumatic muscle finds applications where high initial force and high acceleration are required as in lifting equipment, clamping devices, and gripping systems.

Due to its lightweight and slim design, pneumatic muscle is suited for applications in aviation, mobile technology, car construction, and highly dynamic devices such as cutting units, simulators, and robotics, in general.

Because of the ability of the pneumatic muscle to react quickly, it can also be used as a drive for handling units.

Other areas of application of pneumatic muscle are clean rooms, biomedicine, sewage treatment plants, areas subject to explosion hazards, and woodworking.

Tandem Cylinder

A tandem cylinder is shown in Figure 8.16. It consists of two or more cylinders mounted in-line with pistons connected by a common piston-rod. It is capable of providing amplified force output as compared to a conventional cylinder of the same bore diameter.

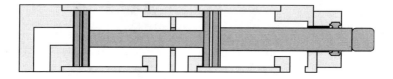

Figure 8.16 | Tandem cylinder

A tandem cylinder can be used where a greater force needs to be accommodated in an actuator of a smaller diameter. It is often used in a place where the radial space in which the cylinder can be located is at a premium, but not the axial space. Applications for this type of cylinder are cushioning on dies, work-holding devices, and lathe equipment.

Multi-position Cylinder

A 3- or 4-position cylinder consists of two separate cylinders of identical diameters, which are interconnected back-to-back by using a multi-position mounting kit. It is to be noted that if one end of the piston-rod of one cylinder is fixed into position, the cylinder barrel will then execute the motion.

A multi-position operation can be obtained depending on the control method used and the way the strokes are subdivided. A 3-position operation can be achieved by using two cylinders of the same stroke lengths, and a 4-position operation can be realised by using two cylinders of different stroke lengths. Multi-position cylinders find wide applications in the machine tool industry where it is used to position materials for successive operations. Figure 8.17 illustrates the operation of a 4-position cylinder.

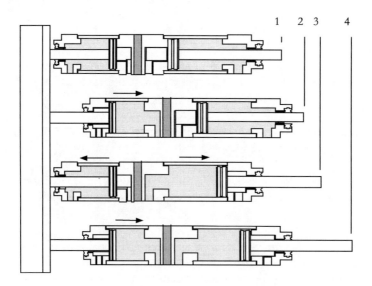

Figure 8.17 | Multi-position cylinder

Impact Cylinder

Pneumatic cylinders are employed for operations requiring huge forces, as in metal forming applications. Pressures and forces that can be developed in pneumatic systems are lower than those in hydraulic systems. However, a huge impact force on a target can be obtained by accelerating the piston of a specially designed pneumatic cylinder to a high velocity and then allowing it to strike the target. Impact cylinders can develop stroke velocities up to 10 m/s and impact energies up to 500 Nm.

The principle of operation of an impact cylinder is illustrated in Figure 8.18. The cylinder is similar to the conventional double-acting cylinder with a provision to store a large amount of compressed air in a reservoir behind the piston. It has two ports A and B.

The cylinder retracts when pressure is applied to port B, as shown in Figure 8.18(a). The pressure is then applied to the exposed small area X of the piston through port A, as shown in Figure 8.18(b). The cylinder remains in the retracted position, as the force acting on area X is less than the force acting on area Y. Port B is then exhausted rapidly using a quick-exhaust valve. As the piston begins to move forward, the full area of the piston experiences the port A pressure, as shown in Figure 8.18(c). With a large volume of compressed air stored behind, the piston quickly accelerates to a high velocity, as shown in Figure 8.18(d).

Impact cylinders can be used as power units capable of providing impact forces to an infinite variety of presswork applications, such as pressing, flanging, riveting, punching, etc. For all applications, complete integrated guarding must be provided, either in a fixed-type configuration or that is interlocked with a control circuit.

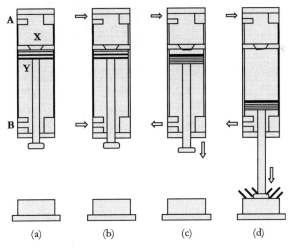

Figure 8.18 | Impact cylinder

Telescopic Cylinder

A telescopic cylinder is an arrangement with nested multiple tubular rod segments that provide a long working stroke in a short restricted envelope. It is used where a longer stroke length than that can be obtained with a standard cylinder, is required. The thrust available from the cylinder is only proportional to the area of the piston of the last stage. The remainder of the piston area is needed to set up the telescopic stages and is ineffective for the generation of the output force. The construction of a telescopic cylinder requires many seals, which makes its maintenance difficult.

Hydro-pneumatic Feed Unit

A conventional pneumatic cylinder is not suitable for an application where uniform working speed is required as in a feed application. Whereas a hydraulic cylinder can deliver uniform speed, it can be used in combination with an easily-controllable pneumatic cylinder to exploit the benefits of both types of cylinders.

A hydro-pneumatic feed unit is shown in Figure 8.19. It consists of a pneumatic cylinder, a hydraulic cylinder with a throttle check valve, and an air control block to form a compact unit. The pneumatic cylinder is engaged as the working element. A crosstie joins the pistons of the pneumatic cylinder and the hydraulic cylinder. An integrated air control block controls the unit.

When compressed air is applied to the pneumatic cylinder, the piston of the pneumatic cylinder moves and carries the hydraulic cylinder along with it. When the hydraulic cylinder moves, oil is transferred from one side of its piston to the other side through the throttle valve. The throttle valve can be adjusted, thereby regulating the feed speed. Here too, the oil prevents the feed from being uneven when the applied load changes.

On the return stroke, the oil can move quickly to the other side of the piston through the check valve (not shown) and hence the return stroke can be made to traverse quickly.

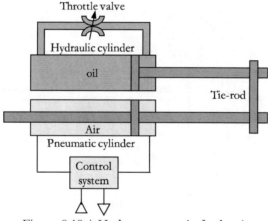

Figure 8.19 | Hydro-pneumatic feed unit

Slow and constant feed movements are possible using the hydro-pneumatic unit. The speed of the working stroke can be regulated typically between 30 to 6000 mm/min.

Rotary indexing table

In many manufacturing processes, it is necessary to carry out feed motions in a circular path. Rotary indexing tables are highly-precise work-positioning devices that index parts to be machined in multiple operations. They are used for rotary indexing work with a high level of indexing accuracy on automatic assembly and packaging machines, embossing and marking equipment, and drilling stations. The powering device in a rotary indexing table is the pneumatic cylinder used in conjunction with an air control block, which controls the movement cycles. The required maximum load torque and resolution of an application are important parameters to be considered while selecting a rotary indexing table. Further, its repeatability and accuracy are to be considered.

Pneumatic Grippers

Modern industrial handling systems and many other automatic systems involve sequential work operations, such as holding, retaining, and subsequent release of work-pieces. Grippers are extensively used to implement these types of work operations. They act as hands in automated machinery. There are two basic types of grippers. They are: (1) Finger-like grippers and (2) Suction grippers

Finger-like Gripper

A finger-like gripper consists of a double-acting pneumatic cylinder with fingers attached to the piston of the cylinder through a rack-and-pinion or lever mechanism. The movement of the piston makes the fingers open or close. The finger shape is to be tailored to the shape and type of surface of the work-piece.

Finger-like grippers are categorised according to: (1) the number of fingers used and (2) the way the fingers are moving. According to the number of fingers, they are classified as 2-point, 3-point, and 4-point grippers. 3-point grippers are preferred for handling cylindrical work-pieces. According to the way, the fingers are moving, grippers are classified as parallel grippers, radial grippers, and angled grippers. For example, in parallel grippers, the fingers move in parallel across the entire stroke. An assortment of different types of grippers is depicted in Figure 8.20.

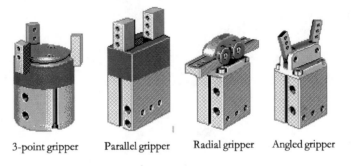

| 3-point gripper | Parallel gripper | Radial gripper | Angled gripper |

Figure 8.20 | Types of grippers

The grippers are usually attached with at least two sensing slots for the sensing of the jaw positions. It is possible to block the exhaust air from the gripper so that the gripped work-piece is stopped from being dropped from the gripper fingers in the event of compressed air failure. Grippers are low-cost devices for providing high gripping force or torque.

Vacuum Grippers

Grippers based on vacuum principles are extensively used in pneumatic systems for picking up delicate parts made of metal, plastic, or wood. A vacuum gripper consists of a vacuum generator and a suction cup.

Vacuum Generator

A vacuum generator, as shown in Figure 8.21 (a), is essentially a 'venturi' with a nozzle and mixer. It also has an inlet port P(1), exhaust port R(3), and vacuum port V(2). A suction cup is connected to the vacuum port. A work-piece with a smooth and non-porous surface can make contact with the suction cup. A vacuum generator produces a vacuum at vacuum port 2 when compressed air flows from port 1 to port 3, through the constriction in the path of airflow, following the venturi principle. The constriction in the venturi nozzle increases the flow velocity of the compressed air to supersonic speed. After exiting the venturi nozzle, the air expands through the mixer and flows into the exhaust port. In this process, the air is drawn in from the vacuum port. If the work-piece blocks the suction cup, a vacuum is produced in the suction cup. The suction cup can then pick up and grip the work-piece.

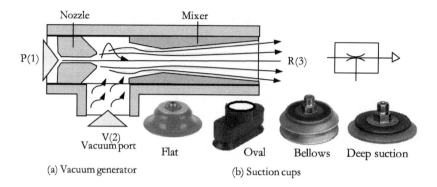

Figure 8.21 | Vacuum generator and Suction cups

If the pressure of 6 bar is applied, typically a vacuum of - 0.8 bar is generated. The surface of a work-piece has to be clean and even to obtain the specified value of suction force. Otherwise, the value is to be reduced depending on the quality of the surface.

Suction Cups

Suction cups are available in a wide variety of sizes with nominal diameters from 0.04 to 25 inches. Figure 8.21(b) shows a few designs of suction cups, which include flat and bellows types in a round or square or oval shape. They are also available in standard/extra-deep designs. Suction cups in the form of bellows allow convex surfaces to be gripped.

The materials used for suction cups are perbunan (Buna-N), polyurethane, silicone, natural rubber, Viton, etc. Work-piece temperature can vary from –50°C to +250°C. At temperatures below 0°C, the hardness of the suction cup may increase making cups virtually rigid and preventing adequate adaptation of the work-piece surface. Polyurethane suction cups can be used in the temperature range from –40°C to +200°C. Silicon suction cups are superior for use in food industries.

Vacuum switch/Vacuum Valve

A vacuum switch (vacuum valve) works as a sensor and checks whether or not the specified vacuum has been achieved. It is connected between the vacuum generator and the suction cup. It is turned on when a particular level of vacuum is generated. When the switch (valve) is triggered, it generates an electrical (pneumatic) signal.

Applications, Vacuum Equipment

The vacuum is extensively used for lifting and handling many types of products and materials, letting many ways of holding, picking up, transporting, and settling down work-pieces. For handling operations, compact vacuum generators with a wide variety of suction cups made of different materials, and vacuum switches (vacuum valves) can be used. The industrial robot is an important handling machine that roughly imitates the human arm with the application of mechanical hands in the form of grippers.

Rotary Actuators, Pneumatic

A pneumatic rotary actuator converts the energy of the compressed air into rotary mechanical energy. Rotary actuators are designed for reciprocating rotary motion or continuous rotation. Accordingly, there are two basic types of rotary actuators in pneumatics. They are semi-rotary actuators and air motors.

Semi-rotary actuators are designed for reciprocating rotary motion up to 360°, and air motors are designed for continuous rotation above 360°.

A rotary actuator can be defined in terms of the torque it produces and its running speed. The starting torque of a rotary actuator is the torque available to move the load from rest. Stall torque is the torque that must be applied by the load to bring a running rotary actuator to rest. Running torque is the torque available at any given speed. The speed, torque, and direction of rotation of rotary actuators can be controlled and adjusted to job requirements.

Rotary actuators are characterised by low installation costs, rapid acceleration, and high starting torque. They are used in all types of applications and in particular, where there is a risk of explosion. They are employed as drives for knives in the papermaking industry, flame-proof drives for mixtures in process industries, or power drives for reciprocating machine tools. They can also be used as prime movers for the continuous movement of scrapers and moving conveyor belts.

Semi-rotary Actuators, Pneumatic

Semi-rotary pneumatic actuators are constructed with a rotating vane or rack-and-pinion design.

The vane type rotary actuator, as shown in Figure 8.22(a), consists of a single vane coupled to the output shaft. It is usually designed for a double-acting operation with a maximum angle of rotation of 270°. Usually, the angle of rotation can be adjusted.

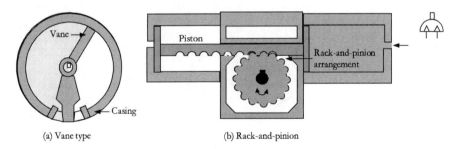

(a) Vane type   (b) Rack-and-pinion

Figure 8.22 | Semi-rotary actuators

The rack-and-pinion type of rotary actuator with limited travel is shown in Figure 8.22(b). It consists of a double-acting piston coupled to the output shaft by a rack-and-pinion arrangement. An angle of rotation up to 360° is possible with this type of design. It can also be designed for double-acting double-torque operation.

The compact rotary actuators are well suited to robotics and material handling applications where there is limited space. They can also be used as a drive for turning components, operating process control valves, and providing a wrist action in robotic applications.

Air Motors

Air motors convert the potential energy of compressed air into rotary mechanical energy. They are designed to provide continuous rotation. Piston, vane, and gear designs are generally used for air motors.

The power, torque, and speed range are the major parameters to be considered in selecting air motors. Typically, air motors are considerably lighter and smaller than electric motors of comparable power ratings. In most cases, air motors are produced in lower power ranges. Typically, air motors are available in the power range from 0.1 to 18 kW. The speed ranges are from 40 to 50000 rpm.

An air motor can accelerate rapidly and withstand repeated stalling and reversing without harm or overheating. However, air motors have disadvantages, which include higher initial costs and lower efficiencies as compared to electric motors.

Applications of air motors are found in powering conveyor belts, printing presses, screwdrivers, hoists and mixers, and many types of portable air tools.

Gear Motor

In the gear motor design, torque is generated by the teeth profiles of two meshed gear wheels. One gear wheel is secured to the motor shaft. Gear motors are available with a very high power rating of up to 45 kW. The applications of gear motors include liquid pump drives, robot arms, mixing, drum pumps, and conveyor drives.

Vane Motor

A rotary vane motor, as shown in Figure 8.23, consists of a cylindrical rotor with sliding vanes placed eccentrically in a cylindrical housing. As air enters the inlet port and passes into the cylinder, a pressure unbalance acts on the vanes. This pressure unbalance develops a torque that turns the rotor against the motor's load. The vane pushes the air out of the exhaust port.

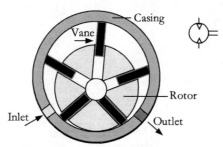

Figure 8.23 | Sliding vane motors

Piston Motors

A piston motor consists of four to six cylinders, which can be arranged in a radial or axial configuration. It develops torque when air pressure is directed to the cylinder pistons. As the pistons reciprocate in sequence, they actuate a wobble plate, which in turn imparts a rotary motion to the output shaft. Piston type motors can arrive at full speed in milliseconds. Piston motors have a high starting torque and good control over their speed. They are well-suited to working with heavy loads at low speeds and for high-quality drive technology and automation. They can be used in winding machines, conveyor belts, positioning machines, and mixers.

Axial Piston Motor

In an axial piston motor, five or more cylinders are arranged axially, and their reciprocating movements are changed into rotary movements through a swash-plate.

Axial piston motors are available in sizes ranging from 0.1 to 2.5 kW and speeds ranging from 220 to 5000 rpm. They are compact, operate smoothly even at low speeds, provide high torque, and reach full speed in a fraction of a second.

Radial Piston Motor

A radial piston motor consists of five or more reciprocating pistons arranged in radial directions. The pistons reciprocate in sequence when compressed air is applied to the pistons. This movement of the pistons causes the associated crankshaft to turn through a connecting rod, thus causing the rotation of the output shaft.

The power output of a radial piston motor is dependent on the operating pressure, the number of pistons, the area of the pistons, the stroke length, and piston speed. Radial piston motors are available in power ratings ranging from 0.1 to 18 kW. The speed of a radial piston motor is limited due to the high inertia of moving mechanical parts. It operates at speeds in the range of 200 to 3000 rpm. Radial piston motors are particularly suited for low-speed operations requiring high starting torque.

Rotary vane motors are available in power ratings ranging from 0.1 to 18 kW and speeds ranging from 800 to 6000 rpm. The efficiency of rotary vane motors is generally low and can be as low as 25%. Rotary vane motors, generally, require lubricated air for optimum sliding efficiency on the housing. They are used for many types of portable tools, power shovels, mechanical saws, lifting jacks, controlling motors, driving pumps, driving blowers, and driving conveyors.

Pneumatic Tools

Air motors are usually used as drives for pneumatic tools. Pneumatic tools are compact, light in weight, portable, and easy to operate. A pneumatic tool is variable in speed, reversible in direction, and can withstand stalling without damage. Speed control can easily be achieved by throttling the air supply to the pneumatic tool.

There are two main classes of pneumatic tools based on their applications. They are portable tools and rock drills. Portable tools carry out a wide range of operations such as nut running, screw driving, grinding, drilling, riveting, scaling, stud driving, and wire wrapping. They include screwdrivers, hammers, riveters, abrasive tools, and hoists. For example, the piston in an air-operated hammer imparts a series of blows to the forming tool fitted at the end of the hammer. Air-operated hoists are used in numerous applications, especially in machine shops and foundries. The air-operated tools include various types of rock drills. The hammer drill is a commonly used air-operated tool in a mining operation or general excavation work.

Air-operated tools can do many difficult tasks easily, quickly, and smoothly. The tasks done by some of the tools are highlighted below: Air wrenches remove bolts and nuts. A pneumatic screwdriver can place screws for industrial jobs. A pneumatic riveter is used to attach rivets. An air-operated paint sprayer can be used for the application of paint on a large surface. A pneumatic die grinder can be used for grinding, sharpening, polishing, and cutting metals. Pneumatic shears are used to cut plastic and metal sheets. A pneumatic sander can be used to make metal or wood surfaces smooth.

Installation of Pneumatic Cylinders

Proper mounting of cylinders onto the machine and coupling of piston-rods are of great importance, as any mismatch will result in stress on the cylinders leading to a reduction in their service life. All mountings must be fastened securely. As soon as force is transmitted to a machine, bearing stresses arise at the cylinder barrel and the piston-rod resulting in high edge pressures on the cylinder bearing bushes and on the piston-rod guide bearings. Increased and uneven stresses may also be developed on the piston seals and piston-rod seals. The different cylinder mounting styles should be studied to determine how they could be installed for the best results.

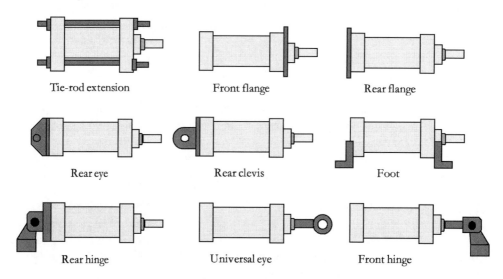

Figure 8.24 | Cylinder mountings

A cylinder should be installed in such a way that side loads on the piston-rod bearing are reduced to a minimum, if not eliminated. A side load is the force component acting laterally across the axis of the bearing, and it can rarely be eliminated. Still, by taking advantage of good engineering practices, they can be reduced to an acceptable limit. Wherever possible, maintain the load on slide or roller guides.

Cylinders can be rigidly fastened to the machine or can be allowed to swivel as part of linkage in one or more planes. The segments to be fixed are the cylinder body and the piston-rod end. Generally used mounting methods are: foot, flange, trunnion, swivel flange, clevis, tie rod, and bolt mounting. The style of mounting is determined by how a cylinder is to be fitted to a machine or fixture. The cylinder can be fixed with a permanent type or an adjustable type of mounting. The latter type can be altered easily as and when required. A few of the essential types of mounting methods are depicted in Figure 8.24.

Standards for Pneumatic Cylinders

ISO 15552, ISO 6432, VDMA 24652, NFE 49003.1, UNI 10290 are various standards for pneumatic cylinders. There are many ISO standards for pneumatic cylinders for specifying their configurations, mounting dimensions, mounting accessories, and testing criteria. The standard ISO 15552:2004 specifies the mounting and accessories dimensions for pneumatic cylinders of bores from 32 mm to 320 mm and pressure ratings up to 10 bar and with detachable mountings. The standard ISO 6432:2015 specifies the mounting dimensions for single-rod pneumatic cylinders of bores from 8 mm to 25 mm and pressure ratings up to 10 bar.

Objective-type Questions

1. A _____ converts compressed air energy into mechanical energy in the form of force and linear movement in one direction of motion only.
   a) piston cylinder
   b) single-acting cylinder
   c) double-acting cylinder
   d) short-stroke cylinder

2. A _____ converts compressed air energy into mechanical energy in the form of force and linear movement in both directions of motion.
   a) single-acting cylinder
   b) double-acting cylinder
   c) single-acting-cylinder, gravity return
   d) single-acting cylinder, spring return

3. What is the theoretical thrust of a 40 mm bore pneumatic cylinder supplied with a pressure of 6 bar?
   a) 242 N
   b) 488 N
   c) 754 N
   d) 2400 N

4. The impact forces of the fast-moving piston and piston-rod assembly of a pneumatic cylinder can be reduced by providing:
   a) magnets
   b) stop tubes
   c) guide rods
   d) air cushioning

5. The cylinder suitable for proximity sensing is provided with:
   a) magnets
   b) stop tubes
   c) guide rods
   d) air cushioning

6. Identify the symbol

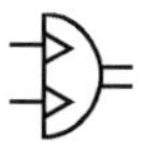

Figure 8.25

   a) Rotary hydraulic actuator
   b) Rotary pneumatic actuator
   c) Semi-rotary hydraulic actuator
   d) Semi-rotary pneumatic actuator

7. Identify the symbol

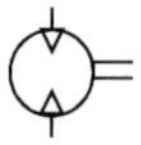

Figure 8.26

a) Rotary hydraulic actuator
b) Rotary pneumatic actuator
c) Semi-rotary hydraulic actuator
d) Semi-rotary pneumatic actuator

8. A linear actuator with equal forces and speeds in both directions of motion
a) tandem cylinder
b) double-acting differential cylinder
c) double-acting cylinder with magnetic pistons
d) double-acting cylinder with through piston-rods on both sides

9. A pneumatic cylinder that provides double the force than that of a standard cylinder
a) ram cylinder
b) impact cylinder
c) tandem cylinder
d) multi-position cylinder

10. A pneumatic cylinder assembly that can be precisely stopped in intermediate locations
a) ram cylinder
b) impact cylinder
c) tandem cylinder
d) multi-position cylinder

11. The type of pneumatic actuator that can accelerate to a great extent with the backup stored energy
a) ram cylinder
b) impact cylinder
c) tandem cylinder
d) multi-position cylinder

12. A pneumatic actuator that can provide uniform motion
a) impact cylinder
b) tandem cylinder
c) multi-position cylinder
d) hydro-pneumatic feed unit

13. A pneumatic rotary unit with a platform that can be positioned precisely at definite angles
a) tandem cylinder
b) rotary indexing table
c) multi-position cylinder
d) hydro-pneumatic feed unit

14. A load of 754 N needs to be pushed upwards using a pneumatic cylinder. What minimum bore size of the cylinder is to be selected if the design pressure is 6 bar?
    a) 20 mm
    b) 32 mm
    c) 40 mm
    d) 50 mm

15. What is the theoretical thrust of a 50 mm bore cylinder having a piston-rod diameter of 32 mm, supplied with compressed air of 6 bar?
    a) 192 N
    b) 482 N
    c) 603 N
    d) 1152 N

16. Mark the <u>incorrect</u> statement:
    a) Pneumatic actuators include cylinders, motors, grippers, and tools
    b) Pneumatic seals for cylinders are made of synthetic rubber, Viton, and PTFE
    c) The maximum bore diameter of standard pneumatic cylinders is 500 mm (20")
    d) Standards for pneumatic cylinders include ISO 15552, ISO 6432, VDMA 24652, NFE 49003.1, and UNI 10290

Questions
1. Define the term actuator and explain its function using a simple double-acting cylinder.
2. Explain the constructional features of a double-acting cylinder.
3. Write the equation for the thrust of a double-acting cylinder.
4. Write the equation for the pull of a double-acting cylinder.
5. Write the equation for the air consumption of a double-acting cylinder for the forward stroke.
6. Write the equation for the air consumption of a double-acting cylinder for the return stroke.
7. Determine the theoretical thrust of a 320 mm bore double-acting pneumatic cylinder supplied with compressed air at a pressure of 7 bar.
8. Calculate the air consumption (V) per cycle of a double-acting cylinder, with a 50 mm bore, 20 mm piston-rod diameter, and stroke length of 500 mm. The cylinder is supplied with compressed air at a pressure of 6 bar.
9. What is cylinder rod buckling? What is the effect of buckling on the cylinder loading?
10. How seals are classified based on application?
11. Differentiate between static and dynamic seals.
12. What are the seals? Through a neat sketch, show the action of a cup seal under pressure.
13. Describe the operation of a 'Z' seal with the help of a neat sketch.
14. State the function of the wiper seal.
15. State the function of the wear ring on the piston of a cylinder.
16. List out different materials used for seals in pneumatic cylinders along with their temperature range.
17. What is the effect of (1) low temperature and (2) high temperature on seals?
18. List out the essential characteristics of cylinder seals.
19. Explain the working principle of a single-acting cylinder with a neat outline sketch.
20. What is the maximum stroke (typical) of a commercially available single-acting cylinder?
21. Differentiate between single-acting and double-acting cylinders.
22. Differentiate between a pneumatic muscle and a conventional single-acting cylinder.

23. Mention the standard values of diameter for pneumatic double-acting cylinders as per ISO 6432.
24. Mention the maximum stroke of a commercial standard double-acting cylinder.
25. How are pneumatic actuators classified?
26. Mention a few applications of pneumatic actuators.
27. List out a few typical applications of single-acting cylinders.
28. Draw the ISO 1219 symbol for a double-acting cylinder with an adjustable cushion.
29. What is the purpose of providing 'cushioning' in pneumatic cylinders?
30. What are the different ways of providing 'cushioning' in pneumatic cylinders?
31. Explain the purpose of providing the following on pneumatic cylinders; (1) Magnet on the piston, (2) Air-cushioning, (3) Integral guide bars, (4) Heat-resistant seals, (4) Piston rod locking unit
32. What are the different designs of rodless pneumatic cylinders?
33. Explain the constructional features of a rodless pneumatic cylinder.
34. Give a few applications of rodless pneumatic cylinders.
35. What are the advantages of rodless pneumatic cylinders?
36. Draw the ISO symbol representing a cylinder with a through-piston rod.
37. Briefly explain the construction of the bellows actuator. What are its applications?
38. What is a pneumatic muscle? Explain its constructional features.
39. Mention a few applications of pneumatic muscle.
40. Explain the design features of a tandem pneumatic cylinder with a neat sketch.
41. What is the advantage of a tandem pneumatic cylinder?
42. Explain the working principle of an impact cylinder. Mention a few applications of impact cylinders.
43. Explain the method of construction of a multi-position cylinder for four positions with a neat sketch.
44. Explain the working principle of the rotary indexing table.
45. What are the two basic types of commercial pneumatic gripper systems?
46. What is the function of pneumatically operated grippers?
47. Explain the working principle of the finger-type gripper.
48. Explain how gripping of work-piece is possible by the arrangement of a vacuum generator and suction
49. Write a brief note on different types of finger-type grippers.
50. Explain the working of a vacuum generator.
51. Mention the different types of suction cups concerning the materials used.
52. What is the function of a suction cup in handling applications?
53. Mention the different types of suction cups concerning the shapes.
54. Mention two typical applications of rotary actuators.
55. How can rotary actuators be classified?
56. What are the advantages of air motors over electric motors?
57. What is the difference between a semi-rotary actuator and an air motor?
58. List five factors that should be considered in selecting a pneumatic cylinder.
59. What is the importance of proper mounting of pneumatic cylinders?
60. What are the precautions that must be taken while installing cylinders?
61. What are the different methods of mounting pneumatic cylinders?
62. Name two ISO standards governing the cylinders and briefly explain them.

[Solutions to problems 7 and 8 are given at the end of the book]

Answer key for objective-type questions:

Chapter 8: 1-c, 2-b, 3-c, 4-d, 5-a, 6-d, 7-b, 8-d, 9-c, 10-d, 11-b, 12-d, 13-b, 14-c, 15-d, 16-c

Chapter 9 | Pneumatic Valves and Control Circuits

The primary purpose of a pneumatic system is to transfer power typically from a compressor or receiver tank to one or more actuators through a compressed air medium for performing some useful work. The path occupied by the fluid from the receiver tank to the actuators and then from the actuators to the exhaust constitutes a pneumatic circuit. The actuator's direction or speed or both are to be controlled during the work process. At times, the system is to be controlled, based on the pressure condition in in a particular part of the circuit. Any simple or complex control requirements can be realised by using an appropriate combination of valves.

What is a Pneumatic Valve?

A pneumatic valve consists of a body with an internal moving element, such as a poppet or spool, actuating mechanisms, and many ports, as shown in Figure 9.1. It is a control device that directs or restricts the compressed air flow or controls the flow based on a specified pressure condition in some part of the associated circuit. Accordingly, pneumatic valves can be classified as directional control valves, flow control valves, and pressure control valves.

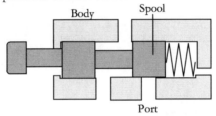

Figure 9.1 | Basic valve construction

Functional Classification of Discrete Hydraulic Valves

The classification of hydraulic valves is given in Figure 9.2. A directional Control (DC) valve (or way valve) controls the path taken by the compressed air. The use of this valve allows compressed airflow in a particular direction for a given switching position. The DC valve can be used to control the direction of the motion of an actuator connected to the system.

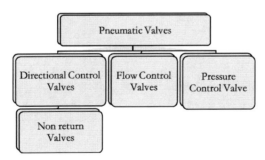

Figure 9.2 | Classification of discrete hydraulic valves

A non-return valve (NRV) allows the flow in a pneumatic system in only one direction and blocks the flow in the opposite direction. A flow control valve restricts the flow rate at which pressurised fluid is transferred in a pneumatic circuit. A pressure sequence valve generates a control signal when the set pressure in some system part has been reached for initiating a subsequent action.

Types of Directional Control Valves
There are two basic types of valves. They are: (1) Poppet valves and (2) Spool/Slide valves.

Poppet Valves
In a poppet valve, a spring-biased disc or ball or cone is used along with the valve seat in a leak-free enclosure to control the compressed airflow. The poppet valve quickly opens a relatively large orifice in short travel to permit the full flow of the fluid. It has the intrinsic characteristic of fast response and finds applications for generating and conveying control signals.

Spool/Slide Valves
In a spool/slide valve, a close-fitting spool moves axially within the body to control the direction of the compressed airflow. The spool valves exhibit excellent shifting characteristics. They are primarily used as main valves to handle the power signals to actuators employed in the system.

Graphic (Symbolic) Representation of Directional Control Valves
Symbols represent pneumatic components because the representations of their complex control functions by sketches are too difficult to draw. The symbolic representation of a component merely specifies the function of the part without indicating its constructional details. The symbols are described in the standard ISO 1219.

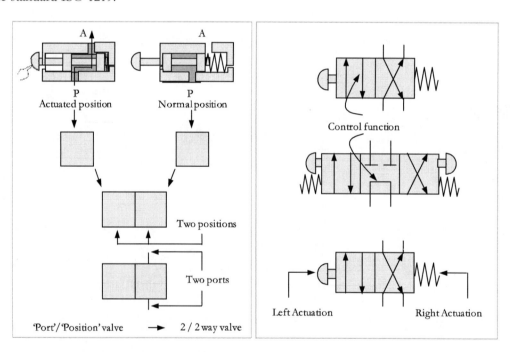

Figure 9.3 | Graphic representation of directional control valves

Figure 9.3 gives the basics of how directional control valves can be represented graphically. A pneumatic DC valve is specified as a 'port/position valve', where the 'port' represents the number of ports, and the 'position' represents the number of switching positions of the valve. Thus, a 3/2-DC valve has three ports and two switching positions. The lines inside the valve represent the function of the valve. The method of actuation of the valve is shown on the left side and right side of the valve.

Symbols for Basic DC Valves

Symbols serve as an aid in the functional identification of the components in the circuit diagrams of fluid power systems. Symbols of basic DC valves are given in Figure 9.4.

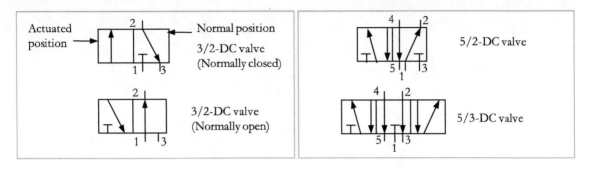

Figure 9.4 | Symbols of basic DC valves

Port Markings

As per the current practice, ports of pneumatic valves are designated using a number system following the ISO 5599 standard. Table 9.1 and Figure 9.5 present the designations for port markings, as per the standard. All the inputs and outputs of DC valves must be identified to avoid faulty connections.

Table 9.1 | Port markings

Port	Letter system (Old system)	Number system (As per ISO 5599)	Comment
Pressure port	P	1	Supply port
Working port	A	2	3/2-DC valve
Working ports	A, B	4, 2	4/2- or 5/2- DC valve
Exhaust port	R	3	3/2-DC valve
Exhaust ports	R, S	5, 3	5/2-DC valve
Pilot port	Z or Y	12	Pilot line (flow 1 -> 2)
Pilot port	Z	14	Pilot line (flow 1 -> 4)
Pilot port	Z or Y	10	Pilot line (flow closed)
Internal pilot ports	Pz, Py	81, 91	Auxiliary pilot air

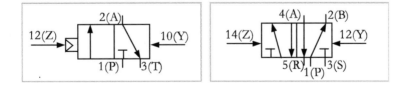

Figure 9.5 | Port markings of DC valves

Methods of Valve Actuation

An essential feature of the directional control valves is the method of their actuation. These valves can be actuated manually, mechanically, pneumatically, or electrically. Figure 9.6 gives the symbols of various actuating methods of DC valves.

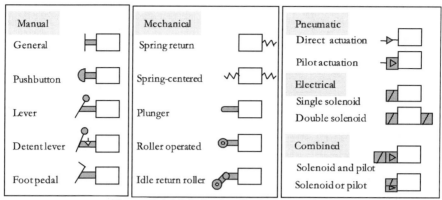

Figure 9.6 | Methods of valve actuation

2/2-DC Valve

Figure 9.7 shows the cross-sectional views of a 2/2-directional (DC) valve in the normal and actuated positions. In the normal position of the valve, pressure port 1 and working port 2 are blocked. In the actuated position of the valve, the working port 2 is open to pressure port 1. Once the actuating force is removed, the spring brings the spool back to its normal position. The 2/2–way valve is capable of blocking or opening the flow passage in a pneumatic system.

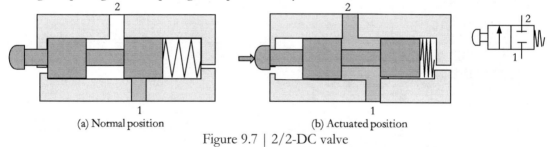

(a) Normal position (b) Actuated position

Figure 9.7 | 2/2-DC valve

2/2-DC valves can be used to switch compressed air supply on and off in main and branch lines, system air supply lines, and airlines to tools and air motors.

3/2-Directional Control Valves

A 3/2-directional control (DC) valve has three controlled connections and two switching positions. This valve can be built with a poppet-type or spool-type design, and with or without an internal pilot valve. There are two versions of 3/2-DC valves, classified according to the way the pressure port is maintained in the normal position. The two versions are: (1) Normally Closed (NC) type and (2) Normally Open (NO) type. In the normally closed 3/2-DC valve, the pressure port is blocked in the normal position. However, in the normally open type, the pressure port is open to the working port in the normal position. The 3/2-way valves can be used to control single-acting cylinders, uni-directional motors, and other valves.

3/2 DC Valves, NC Type

In a normally closed type 3/2-DC valve, the pressure port (1) is closed, and the working port (2) is open to the exhaust port (3) in the normal position of the valve. Further, the pressure port (1) is open to the working port (2), and the exhaust port (3) is blocked in the actuated position of the valve.

3/2 DC Valves – Ball Poppet Design, NC-type

The cross-sectional views of a plunger-operated 3/2-DC valve with ball poppet design in the normal position as well as the actuated position are shown in Figure 9.8.

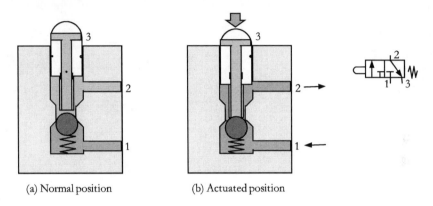

(a) Normal position   (b) Actuated position

Figure 9.8 | 3/2-DC valves (NC) – Ball poppet type

In the normal position of the valve, a spring forces the ball against a valve seat with the result that working port 2 is closed (blocked) to pressure port 1 and open to exhaust port 3. The actuation of the valve plunger against the reset spring and force due to compressed air causes the poppet to be moved away from the valve seat. In the actuated position of the valve, working port 2 is open to pressure port 1 and closed to exhaust port 3.

3/2 DC Valves – Disc Poppet Design (With Internal Pilot), NC-type

The cross-sectional views of a pilot-operated 3/2-DC valve with disc poppet design in the normal position as well as the actuated position are shown in Figure 9.9. The valve consists of a pilot valve and the main valve. The pilot valve is connected to the pressure port via a small hole. In the normal position of the valve, working port 2 is closed to pressure port 1 and open to exhaust port 3.

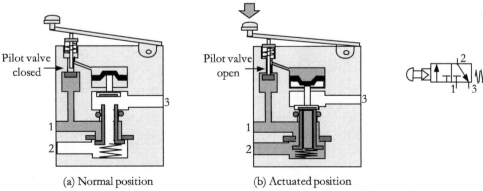

(a) Normal position   (b) Actuated position

Figure 9.9 | 3/2-DC valves (NC) – Disc Poppet type with internal pilot

When the valve is actuated, the pilot valve opens first, causing the compressed air to flow to a diaphragm for the actuation of the main valve. The larger force developed by the compressed air, acting on the diaphragm, moves the disc of the main valve through a plunger, thus actuating the main valve. In the actuated position of the valve, working port 2 is open to pressure port 1 and closed to exhaust port 3. The advantage of a pilot operation is that the necessary actuating force can be reduced.

3/2 DC Spool Valves, NC-type

The cross-sectional views of a 3/2-DC valve (NC type) based on the spool design, in the normal position as well as the actuated position, are shown in the self-explanatory Figure 9.10.

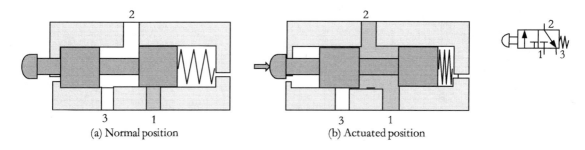

(a) Normal position   (b) Actuated position

Figure 9.10 | 3/2 DC valves, spool design, NC type

3/2 DC Valves, NO-type

In a normally open (NO) type 3/2-DC valve, power supply port 1 is open to working port 2 in the normal position. The cross-sectional views of a 3/2-DC spool valve (NO type), in the normal position as well as the actuated position, are shown in Figure 9.11.

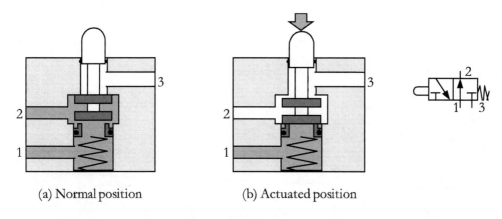

(a) Normal position   (b) Actuated position

Figure 9.11 | 3/2 DC valves, NO type

Example 9.1 | Direct Control of a Single-acting Cylinder

A single-acting cylinder of small-bore diameter is to clamp a component when a pushbutton valve is pressed. As long as the pushbutton valve is pressed, the cylinder is to remain in the clamped position. If the pushbutton valve is released, the cylinder is to retract. Develop a pneumatic control circuit for the control requirement.

Pneumatic circuit

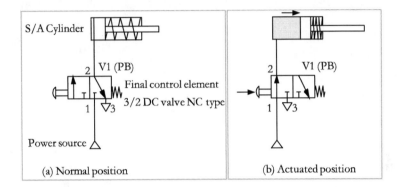

Figure 9.12 | Two positions of the circuit for the control task in Example 9.1

A 3/2-DC normally-closed pushbutton valve controls a small-bore single-acting cylinder. Figure 9.12 shows two positions of the pneumatic circuit in the normal and actuated positions of the valve for controlling the cylinder. Figure 9.12(a) shows the normal position when pushbutton valve V1 is released. In the normal position, the valve blocks the flow from the power source to the cylinder and permits the return flow from the cylinder to the exhaust (2 -> 3). The cylinder retracts to its home position with the spring force. In the actuated position of the vale (when the pushbutton valve is pressed), as shown in Figure 9.12(b), the valve allows the flow from the power source to the cylinder (1 -> 2). The cylinder then extends. In Figure 9.12(b), the arrow indicates the application of a force.

It may be noted that the 3/2-DC valve acts as the final control element that directly controls the single-acting cylinder. The flow rate capacity of the valve should match the system flow rate requirement and hence the size of the cylinder.

Non-return Valves

Non-return valves (NRVs) are devices that preferentially permit the compressed air flow in one direction and stop the flow in the opposite direction. The basic non-return valve is known as a check valve. There are also other derivatives of the basic non-return valve. Check valves can be incorporated as elements in one-way flow control valves, shuttle valves, two-pressure valves, and quick-exhaust valves.

Check Valve

A check valve permits the flow in one direction but stops the flow entirely in the opposite direction. A check valve consists of a valve body and a spring-biased poppet, like a cone, ball, plate, or diaphragm. The schematic diagrams of a check valve in two positions are given in Figure 9.13. When the system pressure at the check valve port 1 is high enough to overcome the low spring force, the poppet is pushed off its seat, allowing air to flow freely through the valve, as shown in Figure 9.13(b).

When the compressed air enters through port 2, the poppet is pressed onto its seat firmly. Hence, the intended flow through the valve gets blocked, as shown in Figure 9.13(a). A check valve is often used as a bypass valve.

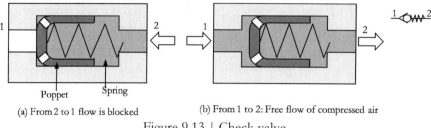

(a) From 2 to 1 flow is blocked

(b) From 1 to 2: Free flow of compressed air

Figure 9.13 | Check valve

Flow-control Valve

In a pneumatic application, it is the flow rate that controls the speed of its actuators. A flow-control valve is used in a pneumatic circuit to control the flow rate of compressed air from one part of the circuit to another part. Its function is to regulate the amount of compressed air passing through the valve using a metering orifice. A throttle valve is an example of a flow control valve. The valve is used to control the speed of a pneumatic actuator in both directions of motion. A one-way flow-control valve can be used to control the speed of the cylinder in a particular direction of its motion.

However, accurate speed control under conditions of changing load is difficult to achieve in a pneumatic circuit. Flow controls can be used satisfactorily in pneumatic applications only when the load does not vary excessively.

Throttle Valve

A throttle valve consists of an opening whose cross-section can be controlled by an externally adjustable needle. The pointed needle can regulate the air flowing through the controlled cross-section. The throttle valve is also called a restriction valve or needle valve. The cross-sectional view of a throttle valve is given in Figure 9.14.

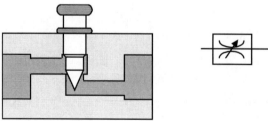

Figure 9.14 | Throttle valve

One-way Flow-control Valve

A one-way flow-control valve is also called a throttle-relief valve or throttle-check valve. This valve is simply a parallel arrangement of a throttle valve and a check valve. The cross-sectional views of a one-way flow-control valve in two positions are given in Figure 9.15. The check valve blocks the flow of air in one direction forcing the air to flow through the restricted cross-section, as shown in Figure 9.15(a). Hence the air is throttled, and hence the flow is controlled in that direction. In the opposite direction, the air flows freely through the opened check valve, as shown in Figure 9.15(b).

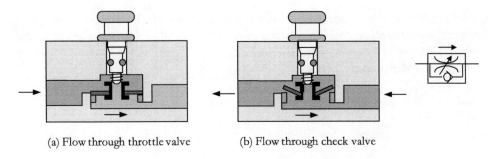

(a) Flow through throttle valve   (b) Flow through check valve

Figure 9.15 | One-way flow control valve (a) Restricted flow through throttle valve (b) Easy flow through the check valve. [Note: The arrow on the valves indicates the direction of throttling.]

Example 9.2 | Speed control of a single-acting cylinder
A single-acting cylinder of a small-bore is to push and clamp jobs when a pushbutton valve is pressed. The forward motion of the piston must be slow. The piston is to retract at normal speed when the pushbutton valve is released. Develop a pneumatic control circuit and designate the ports. The schematic diagram for the control task is given in Figure 9.16.

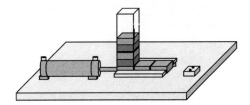

Figure 9.16 | Schematic diagram (Example 9.2)

Pneumatic circuit

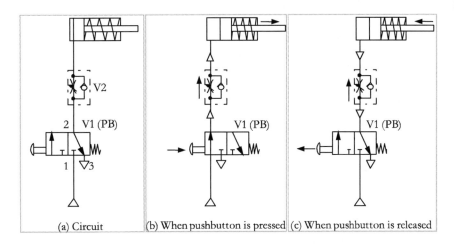

(a) Circuit   (b) When pushbutton is pressed   (c) When pushbutton is released

Figure 9.17 | Three positions of the circuit for the control task in Example 9.2

A 3/2-DC normally-closed pushbutton valve V1 controls the direction of the single-acting pneumatic cylinder, and a throttle-check valve V2 controls the speed of the cylinder. Figure 9.17 shows three positions of the circuit in the normal and actuated positions of valve V1 for controlling the speed of the cylinder. In the normal position, as shown in Figure 9.17(a), valve V1 permits a free flow of compressed air from the cylinder to the exhaust through the opened check valve V2 and blocks the flow from the power source to the cylinder.

In the actuated position, as shown in Figure 9.17(b), valve V1 allows a restricted flow from the power source to the cylinder through the throttle valve V2, as the check valve V2 is closed. The cylinder then extends slowly as per the setting of the throttle valve.

Figure 9.17(c) shows the position of the circuit when the pushbutton is just released. In this position, valve V1 permits a free flow of compressed air from the cylinder to the exhaust through the opened check valve and blocks the flow from the power source to the cylinder. The cylinder then retracts to its home position, as given in Figure 9.17(a).

In a pneumatic system, as the cylinder load varies, air expands or compresses. Consequently, it is impossible to achieve a constant piston-rod velocity within reasonable limits with an ordinary flow-control valve. For this reason, pneumatic cylinders are seldom used in applications where uniform speed must be achieved.

3/2-DC Pneumatically-actuated Valve

The cross-sectional views of a pneumatically-actuated 3/2-DC valve in the normal position and actuated positions of the valve are shown in Figure 9.18.

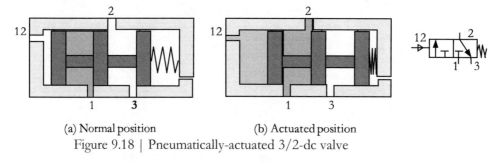

(a) Normal position   (b) Actuated position

Figure 9.18 | Pneumatically-actuated 3/2-dc valve

In the normal position of the valve, as shown in Figure 9.18(a), working port 2 is closed to pressure port 1 and open to exhaust port 3. When the compressed air is directed to pilot port 12, the valve spool is moved against the reset spring. In the actuated position, as shown in Figure 9.18(b), working port 2 is open to pressure port 1 and closed to exhaust port 3.

A pneumatically-actuated valve acts as the final control element to control an actuator in a pure pneumatic power transmission system. The controller in the system generates the control signal to actuate the valve as per the control logic. With the use of the pneumatically-actuated valve, the pneumatic circuit falls into two distinct parts. That is, (1) the main power transmission part and (2) a control part. Many control functions can easily be incorporated into the control part to satisfy application requirements. The flow rate requirement is the main concern to the components in the power transmission system, and hence the size of the final control element should match the flow rate requirement of the actuator. However, the flow rate requirement is of no consequence to components in the pneumatic control part.

Example 9.3 | Indirect control of a single-acting cylinder

A large-bore single-acting pneumatic cylinder, as shown in Figure 9.19, is employed for clamping work-pieces. The forward and return strokes of the cylinder must occur when control signals are given remotely. Develop a pneumatic control circuit.

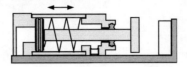

Figure 9.19 | An arrangement for the clamping of work-pieces using a single-acting cylinder

Pneumatic circuit

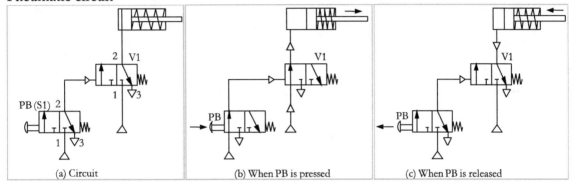

Figure 9.20 | Three positions of the circuit for the control task in Example 9.3

A large-bore single-acting cylinder is controlled by a pneumatically-actuated 3/2-DC valve V1, which acts as the final control element. The control signal for the pneumatically-actuated 3/2-DC valve V1 is generated by a 3/2-DC manually-actuated valve S1. Figure 9.20 shows the critical positions of the circuit for controlling the pneumatic cylinder.

The normal position of the circuit is shown in Figure 9.20(a). Pushbutton valve PB (S1) generates a signal when pressed. The signal is used to control the pneumatically-actuated 3/2-DC valve V1. When actuated, valve V1 permits the compressed air flow from the power source to the cylinder (1 -> 2), as shown in Figure 9.20(b). The cylinder then extends. When the valve PB (S1) is released, as shown in Figure 9.20(c), valve V1 blocks the flow from the power source to the cylinder and permits the flow of compressed air from the cylinder to the exhaust (2 -> 3). The cylinder then retracts to its home position.

Manually Actuated 5/2-DC Valve

A 5/2-DC valve has five ports and two switching positions. The cross-sectional views of a 5/2-DC valve with spool type design are shown in Figure 9.21 in its normal as well as actuated positions.

In the normal position, as shown in Figure 9.21(a), flow paths from port 1 to port 2 of the valve and from port 4 to port 5 are open, and exhaust port 3 is closed. In the actuated position, as shown in Figure 9.21(b), flow paths from port 1 to port 4 and from port 2 to port 3 are open, and exhaust port 5 is closed.

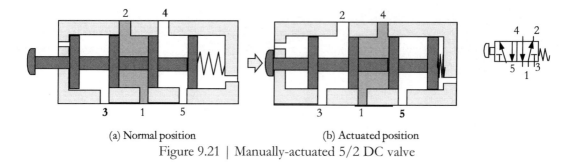

(a) Normal position   (b) Actuated position

Figure 9.21 | Manually-actuated 5/2 DC valve

The typical design of a spool-type directional control valve lends itself readily to a five-port arrangement, which includes two working ports, two exhaust ports, and one pressure port. 5/2-DC valves are used to control double-acting cylinders and bi-directional rotary actuators.

Example 9.4 | Direct control of a double-acting cylinder

A double-acting pneumatic cylinder of a small-bore is to extend and clamp a work-piece when a pushbutton valve is pressed. As long as the pushbutton valve is actuated, the cylinder is to remain in the clamped position. If the pushbutton valve is released, the cylinder is to retract. Develop a pneumatic control circuit.

Solution

A 5/2-DC pushbutton valve V1 controls the small-bore double-acting pneumatic cylinder. The valve acts as the final control element that directly controls the cylinder. Figure 9.22 shows two positions of the pneumatic circuit in the normal and actuated positions of the valve for controlling the cylinder.

Pneumatic circuit

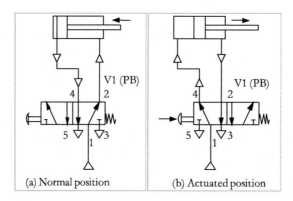

(a) Normal position   (b) Actuated position

Figure 9.22 | Two positions of the circuit for the control task in Example 9.4

In the normal position (when the pushbutton valve V1 is released), as shown in Figure 9.22(a), the valve permits the flow of compressed air from port 1 to port 2 and from port 4 to port 5. Port 3 remains blocked. The cylinder retracts to its home position.

In the actuated position (when the pushbutton valve V1 is pressed), as shown in Figure 9.22(b), the valve permits the flow of compressed air from port 1 to port 4 and from port 2 to port 3. Port 5 remains blocked. The cylinder then extends.

Pneumatically-actuated 5/2-DC Valve

The cross-sectional views of a pneumatically-actuated 5/2-DC valve in the normal and actuated positions of the valve are shown in Figure 9.23.

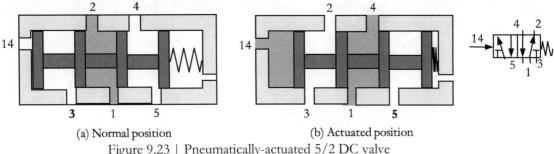

(a) Normal position   (b) Actuated position

Figure 9.23 | Pneumatically-actuated 5/2 DC valve

In the normal position of the valve, as shown in Figure 9.23(a), flow paths from port 1 to port 2 and from port 4 to port 5 are open, and exhaust port 3 is closed. When the compressed air is directed to pilot port 14, as shown in Figure 9.23(b), the valve spool is moved against the reset spring. When the valve is actuated, flow paths from port 1 to port 4 and from port 2 to port 3 are open, and exhaust port 5 is closed.

Example 9.5 | Indirect control of a double-acting cylinder

A large-bore double-acting pneumatic cylinder is employed for clamping work-pieces, as shown in Figure 9.24. The forward and return strokes of the cylinder must occur when control signals are given remotely. Develop a pneumatic control circuit.

Schematic diagram

Figure 9.24 | Clamping of work-pieces

Pneumatic Circuit

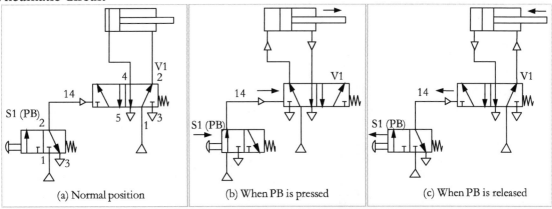

(a) Normal position   (b) When PB is pressed   (c) When PB is released

Figure 9.25 | Three positions of the circuit for the control task in Example 9.5

A large-bore double-acting cylinder is to be controlled by a pneumatically-actuated 5/2-DC valve (V1), which acts as the final control element. The control signal for the pneumatically-actuated 5/2-DC valve is generated by a 3/2-DC manually-actuated valve (S1).

Figure 9.25 shows the critical positions of the circuit for controlling the pneumatic cylinder. The normal position of the valve is shown in Figure 9.25(a). The pushbutton valve S1 generates a signal when it is pressed, as shown in Figure 9.25(b). The signal is used to control the pneumatically-actuated 5/2-DC valve V1. When actuated, valve V1 permits the compressed air flow from port 1 to port 4 and from port 2 to port 3. The cylinder then extends. When valve S1 is released, as shown in Figure 9.25(c), valve V1 permits the flow of compressed air from port 1 to port 2 and from port 4 to port 5. The cylinder then retracts to its home position.

Speed Control of a Double-acting Cylinder

Compressed air is a fast working medium for both linear and rotary motions. Operating speeds of standard cylinders range up to 1.5 m/s. The operating speed of a pneumatic cylinder is decided by how quickly compressed air is filled into the chamber on one side of the cylinder piston and exhausted from the other side of the piston. The speed of the cylinder can be decreased by controlling the rate at which compressed air is allowed in or taken off from the cylinder using flow control valves (throttle valves) installed between the actuator and the main valve.

Speed control is often required to be direction-sensitive, and this necessitates the use of a check valve in parallel with the throttle valve. This parallel arrangement of the check valve and the throttle valve is known as a throttle-check valve or throttle-relief valve. The throttle-relief valve permits the easy flow of air in one direction and the restricted flow of air in the other direction. Similar results can be achieved with the use of merely two throttle valves at the exhaust ports of a five-port main valve.

The throttle-relief valves can be installed for the direction-sensitive speed control of a double-acting pneumatic cylinder in either of the following two ways: (1) Supply-air throttling and (2) Exhaust-air throttling.

Supply-air Throttling

This method of speed control of double-acting cylinders is also called the meter-in method. In the supply air throttling method, the speed of the cylinder in one direction can be controlled by restricting the flow of compressed air into the cylinder and allowing unrestricted flow away from the cylinder. That is, throttle-relief valves are installed between the cylinder and the main valve in such a way that the air entering the cylinder is throttled in each direction of motion of the cylinder. The exhaust air can pass freely through the corresponding check valve in each case.

Exhaust-air Throttling

This method of speed control of double-acting pneumatic cylinders is also called the meter-out method. In the meter-out method, the speed of the cylinder in a given direction can be controlled by restricting the flow away from the cylinder and allowing an unrestricted flow into the cylinder. This method maintains constant backpressure while the cylinder is moving. That means the piston is loaded between two cushions of air while the cylinder is in motion and hence a smooth motion of the cylinder can be obtained. Therefore, exhaust-air throttling is practically used for the speed control of double-acting pneumatic cylinders.

Example 9.6 | Speed control of a double-acting cylinder

An arrangement of a vacuum generator and suction cup is fitted to the piston-rod of a pneumatic double-acting cylinder to pick up and move work-pieces. The speeds of the forward and return strokes are to be regulated. Develop a pneumatic control circuit.

Pneumatic Circuit

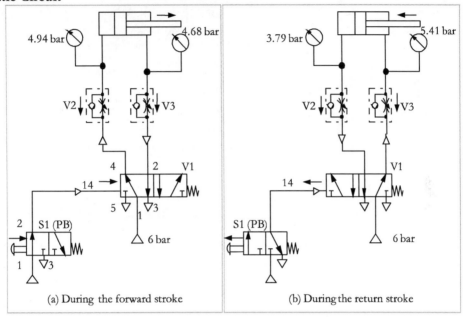

Figure 9.26 | Two positions of the circuit for the control task in Example 9.6

Two positions of the pneumatic circuit for the speed control of a double-acting cylinder in the normal and actuated positions, using the exhaust-air throttling, are shown in Figure 9.26. The throttle-relief valve V3, connected to port 2 of the final control element V1, controls the speed of the forward motion, and the throttle-relief valve V2, connected to port 4 of the final control element V1, controls the speed of return motion. From the Figure, observe the pressure conditions that exist on either side of the cylinder.

5/2-DC Double-pilot Valve

The cross-sectional views of a 5/2-DC double-pilot valve in two positions are depicted in Figure 9.27. A pneumatic signal at pilot port 14 causes the spool to switch over to the right-hand side, as shown in Figure 9.27(b). In this position, flow paths from port 1 to port 4 and from port 2 to port 3 are open, and exhaust port 5 is closed. The valve remains in that position even if the signal to port 14 is removed. If a signal is applied to pilot port 12, the spool switches over to the left-hand side, as shown in Figure 9.27(a). In this position, flow paths from port 1 to port 2 and from port 4 to port 5 are open, and exhaust port 3 is blocked. The valve remains in this position even if the signal at port 12 is removed. Hence, it can be observed that the spool in a double-pilot valve moves to one particular direction when a signal is applied to one pilot port and remains in that position until a pulse or continuous signal is applied to the other pilot port. It may be noted that double pilot valves exhibit memory characteristics.

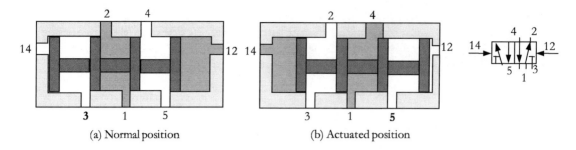

2 4   2 4

(a) Normal position   (b) Actuated position

Figure 9.27 | 5/2-DC double-pilot valve

Application, Double-pilot Valve

The double-pilot valves are found to be convenient valves for realising many control functions easily as these valves are equally reactive to short pulses as well as continuous signals representing system variables. A double-pilot valve is also called an impulse valve or a memory valve as only an impulse is sufficient to switch the valve over.

Signal Conflict

A major problem with a memory valve is the inability to change its switching position when pilot signals appear at both pilot ports of the valve simultaneously. These signals produce equal and opposite forces on the valve spool and hence the spool tends to remain stationary until one of the signals goes off. This problem, called 'signal conflict' or 'signal overlap', is a major hurdle in complex pneumatic circuits. Various control methods have been devised to overcome the problem of signal conflicts, and these are explained in the book 'Pneumatic Systems and Circuits -Advanced Level' by the same author.

Example 9.7 | Memory control of a double-acting cylinder

A double-acting pneumatic cylinder is to extend when a 3/2-way pushbutton valve (PB1) is actuated. This cylinder should remain in the extended position until another 3/2-way pushbutton valve (PB2) is pressed for initiating the return stroke. The cylinder should remain in the retracted position until a new start signal is given using PB1. The speed of the cylinder is to be adjusted in both directions of piston motion. Develop a control circuit to implement the control requirements using a 5/2-way valve as the final control element.

Pneumatic Circuit

Two positions of the pneumatic circuit are shown in Figure 9.28. When PB1 is pressed momentarily (or continuously), a signal is applied to the pilot port 14 of the 5/2-way valve V1. The valve switches over, connecting port 1 to port 4. The piston travels out, as shown in Figure 9.28(a). The piston then remains in the forward end position, until pushbutton PB2 is pressed.

When PB2 is pressed momentarily or continuously, a signal is applied to the pilot port 12 of the 5/2-way valve. The valve switches over, connecting port 1 to port 2. The piston then returns to its initial position, as shown in Figure 9.28(b). The piston then remains in the rear end position, until pushbutton PB1 is pressed again.

This pneumatic circuit is called a memory circuit as the 5/2-way double pilot valve can 'remember' the last signal applied. However, it may be noted that the spool will be unable to move its position when both pushbuttons are pressed simultaneously.

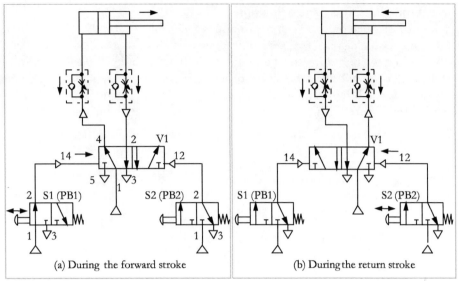

(a) During the forward stroke   (b) During the return stroke

Figure 9.28 | Two positions of the circuit for the control task in Example 9.7

5/3-DC Valves

A 5/3-DC valve has 5 ports and 3 switching positions. 5/3-DC valves are spring-centred valves and can be designed with many types of centre positions. The most important centre positions are: (1) All closed centre position, (2) Open exhaust centre position, and (3) Open pressure centre position. (See Figure 9.29)

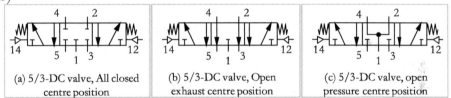

(a) 5/3-DC valve, All closed centre position   (b) 5/3-DC valve, Open exhaust centre position   (c) 5/3-DC valve, open pressure centre position

Figure 9.29 | 5/3-DC valves with various centre positions

5/3-DC Valve, All Closed Centre Position

A 5/3-DC valve, all closed centre position is shown in Figure 9.29(a). The valve can be brought to the left-hand envelop position when a signal is applied to port 14 of the valve. The left-hand envelop position is used to get the forward stroke of the associated double-acting cylinder. The valve can be brought to the right-hand envelop position when a signal is applied to port 12 of the valve. The right-hand envelop position is used to get the return stroke of the cylinder. The spring-centred valve is brought to the centre position when there is no signal applied to the valve. In the centre position of the valve, all ports are blocked. This valve can be used to jog or stop a cylinder in mid-stroke.

5/3-DC Valve, Open Exhaust Centre Position

A 5/3-DC valve with an exhaust centre position is shown in Figure 9.29(b). In the centre position of the valve, all the working ports of the valve are connected to the exhaust. Therefore, the chambers on both sides of the piston of the associated double-acting cylinder remain exhausted when the valve is brought to its centre position. As the piston is floating, the cylinder cannot hold its position against external loads. Figure 9.30(a) shows a self-explanatory circuit with an open exhaust centre position.

5/3-DC Valve, Open Pressure Centre Position

A 5/3-DC valve with an open pressure centre position is shown in Figure 5.29(c). In the centre position of the valve, equal pressure can be applied to both sides of the piston of the associated double-acting cylinder. If the cylinder is a non-differential cylinder, the piston can hold its position, when the valve is brought to its centre position. If the cylinder is a differential cylinder, both sides of the piston create an unequal force, due to area differential, which causes the cylinders to extend when the valve is brought to its centre position. Figure 9.30(b) shows self-explanatory circuits with the open pressure centre position.

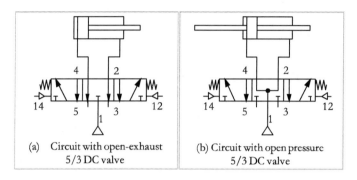

(a) Circuit with open-exhaust 5/3 DC valve

(b) Circuit with open pressure 5/3 DC valve

Figure 9.30 | Typical pneumatic circuits with 5/3-DC valves

Logic Controls, Pneumatic

Many work operations in an industrial pneumatic system are carried out upon fulfilling various logic conditions in the system. The state of a condition is binary and may be regarded as true (1) or false (0). A logic device/element consists of many inputs and usually one output. Next, the logic device produces an output signal depending on the input signals. There are many standard logic functions recognised in control systems. The most important logic functions are 'OR' logic and 'AND' logic.

The OR logic valve has two inputs X and Y and one output A. The output is present if one or more of the input signals are present. The output is absent only if all the input signals are absent. All combinations of the inputs and the output can be shown in a table called the 'truth table'. A two-input OR valve and its truth table are given in Figure 9.31(a).

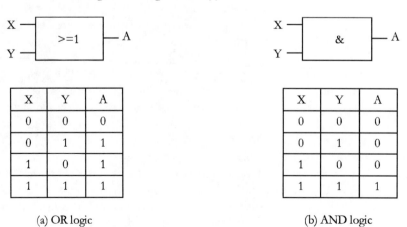

X	Y	A
0	0	0
0	1	1
1	0	1
1	1	1

X	Y	A
0	0	0
0	1	0
1	0	0
1	1	1

(a) OR logic

(b) AND logic

Figure 9.31 | Logic functions

The AND logic valve has two inputs X and Y and one output A. The output is present only if all the inputs are present, and the output is absent if one or more of the inputs are absent. A two-input AND valve and its truth table are given in Figure 9.31(b).

Logic valves for OR operation and AND operation are extensively used in pneumatics. The pneumatic valves used to implement logic functions are shuttle valves and two-pressure valves.

Shuttle Valve

The cross-sectional views of a shuttle valve in two switching positions are given in Figure 9.32. The valve has two inputs 12 and 14, and one output 2. A disk poppet moves inside. If the compressed air is applied to input 12, the disk seals off input 14, and compressed air flows from input 12 to output 2, as shown in Figure 9.32(a). If the compressed air is applied to input 14, the disk seals off input 12, and compressed air flows from input 14 to output 2, as shown in Figure 9.32(b). When input signals are applied to both the inputs of the valve, the signal that is applied first flows to the output. Therefore, the valve realises the OR logic function.

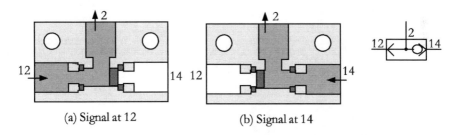

(a) Signal at 12 (b) Signal at 14

Figure 9.32 | Two critical positions of a shuttle valve

Two-pressure Valve

The cross-sectional views of a two-pressure valve in two switching positions are given in Figure 9.33. The valve has two inputs (12 and 14) and one output (2). If compressed air is applied to either input 12 or input 14, the spool moves to block the flow, and no signal appears at output 2 [Figure 9.33(a)]. If signals are applied to inputs 12 and 14 one after another, then the signal applied last flows to output 2. Therefore, the valve realises the AND logic function, as shown in Figure 9.33(b).

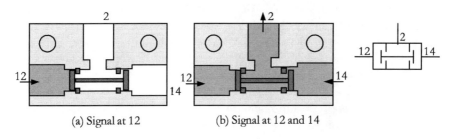

(a) Signal at 12 (b) Signal at 12 and 14

Figure 9.33 | Two critical positions of a two-pressure valve

Extension of the Number of Inputs, Logic Valves

A pneumatic AND or OR valve is designed with only two inputs, as a standard. Many logic valves will have to be interconnected to extend the number of inputs.

Application, Logic Valves

A shuttle valve (OR valve) can be used for the logic control of a machine from different locations. A two-pressure valve (AND valve) can be used as a safety interlock in a machine or system as many conditions must be satisfied for its safe operation. For example, a machine can be designed with interlocks to enable it only when safety guards are in place or disable it in the event of an unsafe condition like an open access panel.

Example 9.8 | Control of a double-acting cylinder with OR logic

A large-bore double-acting cylinder is to extend and clamp a work-piece when either a pushbutton valve or a foot-pedal valve is pressed. The cylinder is to retract when a second pushbutton is pressed. Develop a pneumatic circuit to implement the given control task.

Pneumatic Circuit

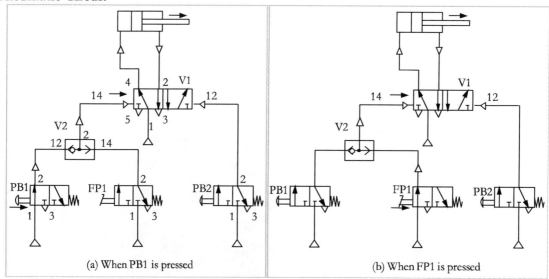

Figure 9.34 | Two positions of the circuit for the control task in Example 9.8

The pneumatic circuit for the control task of Example 9.8 for the control of a double-acting cylinder is given in Figure 9.34. The circuit uses a double-pilot 5/2-DC valve V1 for the directional control of the cylinder, a shuttle valve V2 for realising the OR logic control involved in the control task, pushbutton valves PB1 and PB2, and a foot-pedal valve FP1. The double-acting cylinder should extend when pushbutton valve PB1 or foot-pedal valve FP1 is pressed (OR logic) and should retract when pushbutton valve PB2 is pressed.

When pushbutton valve PB1 is pressed momentarily or continuously, compressed air flows from port 1 to port 2 of the valve, and a control signal is generated momentarily or continuously, as the case may be. This signal is applied to port 12 of the shuttle valve V2, which, in turn, generates an output signal through port 2. The output signal of valve V2 is then applied to the 5/2-DC valve V1. Therefore, valve V1 is actuated for the left envelop for realising the forward stroke of the cylinder.

Similar actions takes place when FP1 is pressed. Here, the generated signal is applied to port 14 of the shuttle valve V2, which, in turn, generates an output signal through port 2.

110

Example 9.9 | Bending machine with two-hand safety
A large-bore double-acting cylinder is to extend and bend a sheet metal when two distantly installed pushbutton valves (PB1 and PB2) are pressed simultaneously. The pushbuttons are installed for two-hand safety which means that both hands of the operator must be engaged to operate the pushbuttons. The cylinder must retract when any one or both pushbuttons are released. Develop a pneumatic control circuit to implement the control task.

Pneumatic Circuit

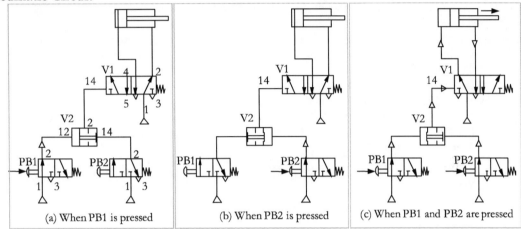

(a) When PB1 is pressed   (b) When PB2 is pressed   (c) When PB1 and PB2 are pressed

Figure 9.35 | Three positions of the circuit for the control task in Example 9.9

The pneumatic circuit for the control task of Example 9.9 for the control of a double-acting cylinder is given in Figure 9.35. The circuit uses a 5/2-DC, spring-return valve V1 for the directional control of the cylinder, a two-pressure valve V2 for realising the AND logic control involved in the control task, and pushbutton valves PB1 and PB2.

The double-acting cylinder should extend when pushbutton valves PB1 and PB2 are pressed (AND logic) and should retract when anyone or both pushbutton valves are released.

Pushbutton valves PB1 and PB2 are connected to the input side of the two-pressure valve V2. The output of two-pressure valve V2 controls valve V1, which in turn controls the double-acting cylinder.

Two-pressure valve V2 produces no output signal if only one pushbutton valve is pressed. Therefore, valve V1 remains in its normal position, and the cylinder remains in its retracted position. The position of the circuit, when pushbutton valve PB1 is pressed, is shown in Figure 9.35(a) and the position of the circuit, when pushbutton valve PB2 is pressed, is shown in Figure 9.35(b).

If both pushbutton valves PB1 and PB2 are pressed, as shown in Figure 9.35(c), two-pressure valve V2 generates an output signal that actuates valve V1, and hence the cylinder extends to realise the AND logic function. If any of the pushbutton valves is released, the circuit returns to its normal position.

Tandem Connection of Valves

An AND function can also be realised through the tandem connection of signal valves, as shown in Figure 9.36, which is self-explanatory. This method of connection for the AND function is economical because of the fewer valves used. However, the disadvantage is that the signal from pushbutton valve PB1 alone cannot be taken out, as this signal is 'AND-ed' with the signal from the pushbutton valve PB2.

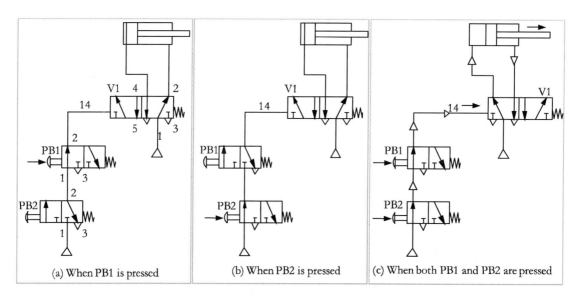

Figure 9.36 | Three positions of a circuit with a tandem connection of valves

Structure of Pneumatic Circuits

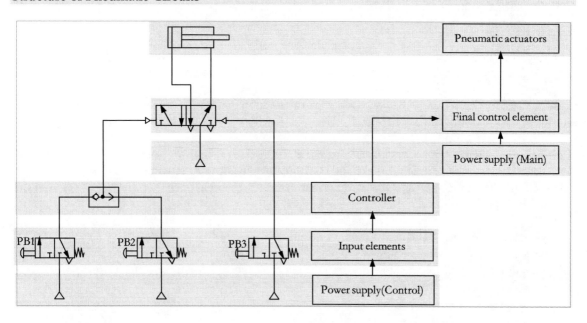

Figure 9.37 | A typical pneumatic circuit and a general structure of pneumatic circuits

A typical pneumatic circuit can be divided into the control part and the power part, as shown in Figure 9.37. The control part consists of a power supply, input elements, and a controller. The signal flow is represented from the bottom to the top as a convention. Signal input elements consist of pushbuttons and sensors, which are used to input signals to the controller. The controller consists of directional control valves, logic valves, timers, counters, etc., which are used to process the received signals and generates outputs to control final control elements.

The power part consists of the main power supply, final control elements, and actuating devices. The final control elements receive signals from the controller and generate signal outputs to control the actuating devices. Actuating devices convert pneumatic energy to mechanical energy in a controlled manner.

Designation of Pneumatic System Components Using Numbers

The components in the pneumatic circuit can be divided into groups such as energy supply group and actuator (working) groups. A pneumatic circuit generally consists of many actuators and control components. An actuator, along with all the associated control components, forms a working group.

The general structure for the designation of a component in a pneumatic circuit is X.Y, where X represents the designation, such as 0, 1, 2, 3, etc., indicating the group to which it belongs, and Y represents the designation, such as 0, 1, 2, 3, 4, 5, etc., indicating the type of component within a group such as an actuator, final control element, or processing and signal valves. The pneumatic system components shown in Figure 9.38 are designated using this method. Table 9.2 gives the designation of pneumatic system components using numbers.

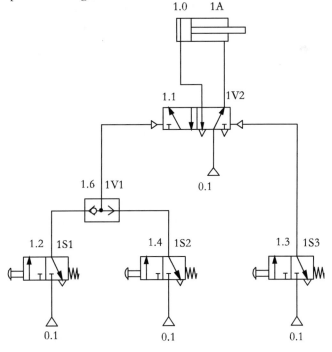

Figure 9.38 | Designation of components in a pneumatic circuit

113

Table 9.2 | Designation of pneumatic system components using numbers

Element/Designation	X	.	Y	Example (X.Y)
Energy supply	0	.	1, 2, 3 ….	0.1, 0.2, 0.3 …….
Working groups (One working group per actuator)				
Actuator			0	1.0, 2.0, 3.0 ……
Final Control Element			1	1.1, 2.1, 3.1 ……
Valves controlling forward motion	1, 2, 3...	.	(2, 4, 6...)	1.2, 2.4, (even numbers for Y)
Valves controlling return motion			(3, 5, 7...)	1.3, 2.5, (odd numbers for Y)
Speed control valve, forward motion			02	1.02, 2.02 ……
Speed control valve, return motion			01	1.01, 2.01 ……

Designation of Pneumatic System Components Using Letters

Alternatively, pneumatic components can be designated using letters. Table 9.3 gives the designation of pneumatic components using numbers and letters.

Table: 9.3 | Designation of pneumatic system components using letters

Element	Designations
Energy supply elements	0Z1, 0Z2…
Actuators	1A, 2A…
Control elements	1V1, 1V2…
Signal elements for the return stroke of cylinders	1S1, 1S2, 2S1…

Representation of Valves Actuated in the Initial Position

A valve, which is actuated in the initial position, must be indicated in a circuit diagram with a trip cam alongside. A 3/2-DC normally-closed type roller valve, which is actuated in the initial position, is drawn in the circuit diagram in a manner, as shown in Figure 9.39.

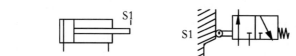

Figure 9.39 | Representation of valves actuated in the initial position

Automatic Control

Pneumatic systems are found to be quite amenable to automation. Sensors play a pivotal role in the automation of industrial production or process systems. Therefore, with the increasing popularity of automatic systems, the study of sensors has become significant for a control engineer.

A sensor is a device for converting a physical variable, like distance, pressure, temperature, etc., of a production or process system into an electrical or pneumatic signal. A sensor can detect whether a particular operation in the system has been completed or not, and then it can generate an output signal to indicate this detection. This signal can be fed back to the associated system controller for triggering the start of the next control action.

Sensors are operated through either physical contact or without contact. Roller valves, limit switches, etc. are examples of contact-type sensors. Magnetic sensors, optical sensors, etc. are some examples of contactless types of sensors.

Roller Valve

A roller valve consists of a 3/2-DC valve fitted with a roller head either directly to the end of the valve plunger or preferably to a lever that operates the plunger. The valve with a roller head is shown in Figure 9.40. On completion of the particular machine operation, the actuator or some moving part of the system actuates the roller valve generating a control signal.

Figure 9.40 | Roller valve

Example 9.10 | Automatic return motion of a double-acting cylinder
A double-acting cylinder is to extend and push work-pieces onto a bin when a pushbutton valve is operated momentarily. After reaching the forward end position, the cylinder is to retract automatically. The arrangement for pushing the work-piece is given in Figure 9.41. Develop a pneumatic control circuit.

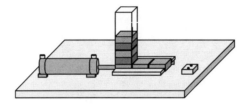

Figure 9.41 | Arrangement for pushing work-piece

Pneumatic circuit

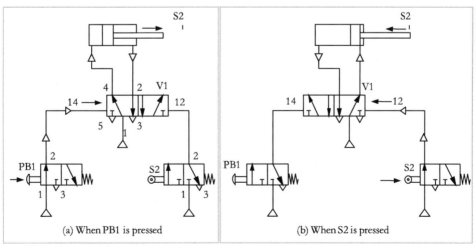

Figure 9.42 | Two positions of a circuit for the automatic return motion of a double-acting cylinder

The double-acting cylinder is controlled by the final control element V1. When PB1 is pressed momentarily, valve V1 is actuated for the left-hand envelop position and the cylinder extends, as shown in Figure 9.42(a). At the forward end position of the cylinder, sensor S2 is actuated automatically by the cylinder. Sensor S2 is pressed, and valve V1 is actuated for the right-hand envelop position, as shown in Figure 9.42(b). The cylinder then retracts automatically. Note, the presence of an operator is essential for the initiation of every cycle. This type of control is known as semi-automatic control.

Quick-exhaust Valve

For many applications, such as punching, bending, etc., it is necessary to get large impact forces. Fast-moving pneumatic cylinders can realise such huge forces. A cylinder in a pneumatic system can move faster than its normal speed if the return air from the cylinder is exhausted quickly through a quick-exhaust valve. The valve is provided with a larger flow path for the return air. Moreover, the valve is usually attached directly to the cylinder ports to avoid the dead volume of air between the cylinder and the main valve controlling the cylinder.

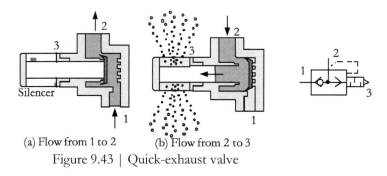

(a) Flow from 1 to 2   (b) Flow from 2 to 3

Figure 9.43 | Quick-exhaust valve

The cross-sectional views of a quick-exhaust valve are given in Figure 9.43. This valve consists of an inlet port 1, an outlet port 2, and a large exhaust port 3. A poppet lip-seal moves inside the valve when pressure acts on it to block either the inlet port or the exhaust port. The normal flow is from port 1 to port 2, as shown in Figure 9.43(a). The quick exhaust flow is from port 2 to port 3, as shown in Figure 9.43(b).

Silencers

Compressed air makes considerable noise when exhausted rapidly to the atmosphere through a pneumatic control valve. A silencer, as shown in Figure 9.44, can be screwed into the exhaust ports of the valve to reduce the exhaust noise within the allowed level.

Figure 9.44 | Silencer

A silencer consists of some damping material and a perforated support cover. The damping material may be made of sintered plastic or sintered metal (aluminium or steel). The exhaust air entering the silencer is distributed over a large surface area of the damping material. The speed of the air is reduced while it is passing through it. This slower pressure relief causes reduced noise. Silencers are available in different sizes ranging from 15 mm to 55 mm. Silencers are not intended to throttle the exhaust air.

Example 9.11 | Rapid motion of a double-acting cylinder

The forming tool on a bending device is driven by a pneumatic cylinder to fold the edge of a flat sheet, as shown in Figure 9.45. The cylinder moves rapidly in the forward direction when two pushbutton valves (PB1) and (PB2) are pressed simultaneously for two-hand safety. If either of the two pushbuttons is released, the double acting cylinder is to return to the initial position. Develop a pneumatic control circuit.

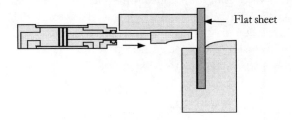

Figure 9.45 | Bending device

Pneumatic circuit

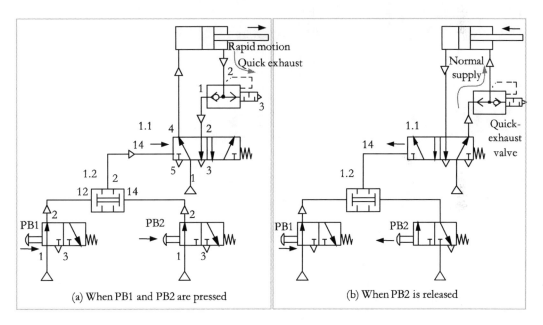

Figure 9.46 | Two positions of the circuit for the control task in Example 9.11

When PB1 and PB2 are pressed simultaneously, as shown in Figure 9.46(a), valve 1.1 is actuated, directing the supply air to the piston-side port of the cylinder through port 4 of valve 1.1. The two-hand safety feature with the AND logic function can be realised by installing two distantly installed pushbutton valves PB1 and PB2.

The return air from the cylinder forces the internal poppet of the quick-exhaust valve to shift and close its port 1, as shown in Figure 9.46(a). The return air is then rapidly relieved through the large exhaust port 3 of the quick-exhaust valve, resulting in the rapid movement of the cylinder.

When PB2 (or PB1) is released, valve 1.1 returns to the normal switching position, as shown in Figure 9.46(b). The supply air flows from valve 1.1 and shifts the internal poppet, to seal its exhaust port 3. This control action permits the supply air to flow through the quick-exhaust valve to the cylinder in the normal way. The cylinder then retracts at the normal speed.

Time-delay Valves

In many pneumatic applications, timing functions are required in the processing of work operations. The necessary timing functions can be achieved using pneumatic time delay valves (timers). A timer consists of a pneumatically actuated 3/2-DC valve, air reservoir, and throttle relief valve. The 3/2-DC valve can be of the normally closed (NC) type or normally open (NO) type. Accordingly, timers are classified into: (1) Timer (NO) type and (2) Timer (NC) type. A normally-closed type timer is shown in Figure 9.47.

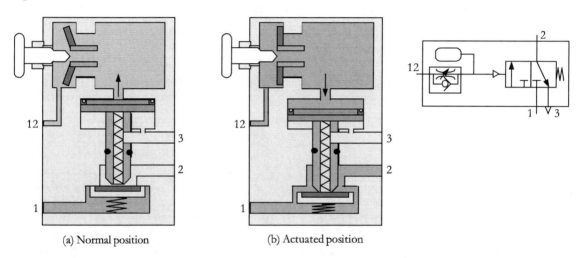

(a) Normal position   (b) Actuated position

Figure 9.47 | Pneumatic timer, normally closed type

In the normal position of the timer, port 12 is connected to the exhaust. Then, the pressure in the reservoir is relieved through the check valve. Also, port 1 of the 3/2-DC valve remains closed. In the actuated position of the timer, port 12 is applied with a pilot signal. Then, the pressure in the reservoir builds up through the throttle valve. Adjustment of the throttle valve permits fine control of the time delay between the minimum and the maximum limits. When sufficient pressure is developed in the reservoir, the 3.2-DC valve is actuated connecting port 1 to port 2 of the valve. In pneumatic time delay valves, typical time delays in the range of 5 to 30s are possible.

A time delay in the operation of the 3/2-DC valve can be effected with reference to the application or release of the compressed air as a pilot signal. Accordingly, pneumatic timers can be classified as: (1) On-delay timer and (2) Off-delay timer.

In the on-delay timer, the 3/2-DC valve is actuated after a delay with reference to the application of the pilot signal to port 12. The valve is reset immediately after the release of the pilot signal. In the off-delay timer, the 3/2-DC valve is actuated immediately on the application of the pilot signal. The valve is reset only after a delay with reference to the release of the pilot signal.

Time-delay Valve, NO-type

The function of a normally-open (NO) type timer is similar to that of a normally-closed (NC) timer, except for the type of 3/2-DC valve used. In the normally-open type timer, a normally-open type 3/2-DC valve is used whereas in the normally-closed type timer, a normally-closed type 3/2-DC valve is used. The symbol of an on-delay timer (NO-type) is shown in Figure 9.48.

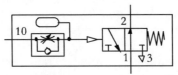

Figure 9.48 | Symbol of a NO-type on-delay timer

Example 9.12 | Pneumatic circuit with time delay

A double-acting cylinder is used to press bonded sheets together, as shown in Figure 9.49. The cylinder is to extend when a pushbutton valve is pressed even momentarily, and then to press the sheets for 5 seconds. The cylinder is to retract after the completion of the pressing operation automatically. Develop a pneumatic control circuit to implement the control task.

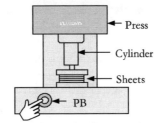

Figure 9.49 | Schematic diagram (Example 9.12)

Pneumatic circuit

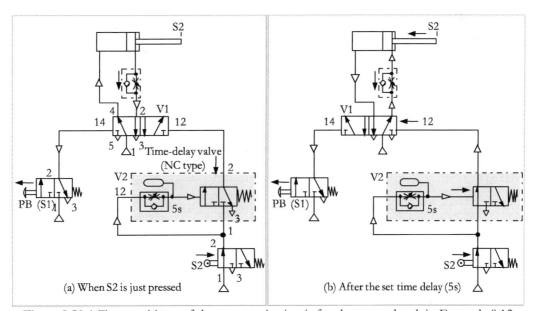

Figure 9.50 | Two positions of the pneumatic circuit for the control task in Example 9.12

119

A 5/2-DC double-pilot valve V1 controls the cylinder. The 3/2-DC pushbutton valve S1 controls the forward motion of the cylinder, and sensor S2 controls the return motion of the cylinder automatically. The required time for the pressing operation is set on a normally-closed type time-delay valve V2.

The cylinder extends when pushbutton valve S1 is pressed and then presses the sheets. Sensor S2 senses the required position of the cylinder and generates a signal. The signal is then directed to the pilot port of timer V2, as shown in Figure 9.50(a). After the set time has elapsed, the timer directs a signal to DC valve V1 for the return motion of the cylinder, as shown in Figure 9.50(b).

Continuous Back and Forth Motion of a Cylinder

A pneumatic cylinder can perform continuous back-and-forth motions automatically with the installation of two sensors in position for sensing the retracted and extended positions of the cylinder. The circuit for the fully automatic operation of a cylinder with a start and stop control (Example 9.13) is given in Figure 9.52.

Example 9.13 | Continuous Back and Forth motion of a double-acting cylinder
A double-acting cylinder is to carry out an oscillatory motion after a 'start' signal is given. The cylinder should stop in the retracted position always when a 'stop' signal is given. Develop a pneumatic control circuit to implement the control task.

Solution
A 5/2-DC double-pilot valve V1 controls the forward and return motions of the double-acting cylinder. The retracted and extended positions of the cylinder are sensed by two sensors S1 and S2 respectively.

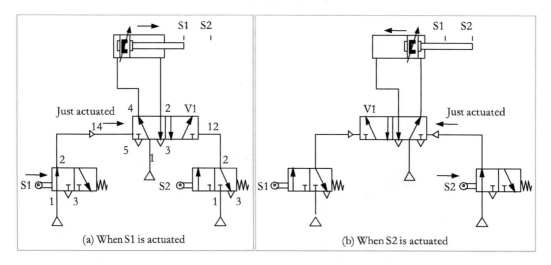

(a) When S1 is actuated   (b) When S2 is actuated

Figure 9.51 | Two positions of the initial circuit for the control task in Example 9.13

When the cylinder is in its fully retracted position, sensor S1 generates a signal automatically, as shown in Figure 9.51(a). The signal is applied to port 14 of valve V1. The valve shifts to its left envelop position to connect port 1 to port 4 of the valve. The flow is then directed to the piston side of the cylinder to move its piston to the forward end position. As soon as the cylinder starts moving forward, sensor S1 drops its signal.

When the cylinder is in its fully extended position, sensor S2 generates a signal automatically. The signal is applied to port 12 of valve V1. The valve shifts to its right envelop position to connect port 1 to port 2 of the valve, as shown in Figure 9.51(b). The flow is then directed to the piston-rod side of the cylinder to move its piston to the retracted position. As soon as the cylinder starts retracting, sensor S2 drops its signal.

The cylinder extends and retracts automatically with the signals generated from sensors S1 and S2. However, the cyclic operation of the cylinder is uncontrollable. Therefore, it is necessary to incorporate the start and stop controls into the control scheme.

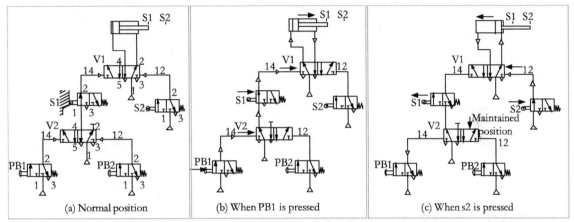

Figure 9.52 | Three positions of the final circuit for the control task in Example 9.13

A 3/2-DC double-pilot valve can be added to the circuit in series with sensor S1 to control the air supply to port 14 of valve V1. However, if a 3/2-DC double-pilot valve is not available, a 5/2-DC double-pilot valve with its port 2 blocked can be used, as shown in Figure 9.52. A 'start' pushbutton valve PB1 and a 'stop' pushbutton valve PB2 can be used to control the switching positions of the double-pilot valve V2.

When a momentary input signal is given to port 14 of 5/2-DC valve V2 through pushbutton valve PB1, as shown in Figure 9.52(b), the valve shifts to its left envelop position and provides a continuous air supply to sensor S1. The cylinder then oscillates with the signals from sensors S1 and S2. The position of the circuit, with sensor S2 pressed, is shown in Figure 9.52(c).

The air supply from 5/2-DC valve V2 goes off only when a momentary input signal is given to port 12 of the valve through the 'stop' pushbutton valve PB2. The cyclic operation of the cylinder then stops when the piston reaches the rear end position.

Pneumatic Preset Counters

A preset counter, in general, performs counting tasks and can count in ascending or descending order. Accordingly, there are two types of pneumatic counters. They are: (1) Up counter, and (2) Down counter. If the set value is counted, the counter generates a pneumatic output signal. The symbol of a down counter is shown in Figure 9.53. An up counter counts in ascending order (typically from a specified number) up to a maximum limit (say 99999). A down counter counts in descending order from a specified number up to zero.

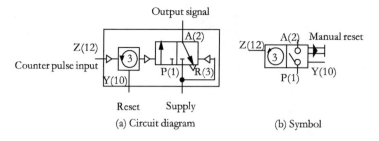

Output signal

Z(12)   A(2)   Manual reset

Counter pulse input

A(2)
P(1)   R(3)   Z(12)   A(2)
Y(10)   P(1)   Y(10)

Reset   Supply

(a) Circuit diagram   (b) Symbol

Figure 9.53 | Symbol of a pneumatic preset counter

A pneumatic preset counter typically consists of the following four ports: P(1) for pressure connection, A(2) for the output signal, Z(12) for the 'count' signal, and Y(10) for the reset signal. It also consists of a repeat button, a reset key, digit keys, and a display window. The repeat button can be used to reset the counter and the digit keys to set the digits.

A counter receives pneumatic signals through port 12 and counts the number of signals received. If the preset number is reached, the counter generates a pneumatic output signal through port 2. The counter can be reset manually using the repeat button or by applying a pneumatic reset signal to port 10.

Example 9.14 | Continuous cycle of operation with counter
A double-acting cylinder is to carry out a continuous back-and-forth motion after a 'start' signal is given. The cylinder should stop automatically after, say, 8 cycles of operation. Develop a pneumatic control circuit to implement the control task.

Pneumatic circuit

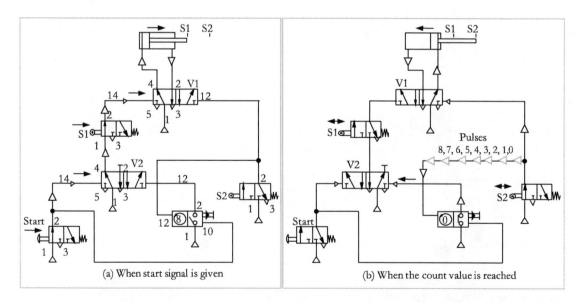

(a) When start signal is given   (b) When the count value is reached

Figure 9.54 | Two positions of the circuit for the control task of Example 9.14

Example 9.14 explains a circuit for the continuous cycle of operation with 'start and stop' controls. Figure 9.54 gives the circuit incorporating a down counter, which is preset with a count value of 8, for counting 8 cyclic work operations. When the start signal is given, as shown in Figure 9.54(a), the system begins the cyclic operations as well as loads the counter with the preset count valve. The count signal is usually taken from sensor S2. Every time sensor S2 senses, the count value is decremented by 1. When the count value becomes 0, as shown in Figure 9.54(b), the counter generates a signal which automatically resets valve V2 and stops the cyclic operations.

Pressure Sequence Valve

There are many pneumatic applications where sequential work operations or a consistent pressing/clamping pressure is essential. Such a requirement can be realised by using pressure sequence valves.

The cross-sectional views of a pressure sequence valve are shown in Figure 9.55. This valve consists of a regulator and pneumatically actuated 3/2-DC valve. Pressure can be set on the regulator. When the preset pilot pressure is reached within the valve due to the building up of the line pressure after the associated cylinder piston reaches the end of its stroke, the spring-loaded piston in the regulator is unseated. The resulting pilot pressure actuates the integrated 3/2-DC valve and generates an output signal.

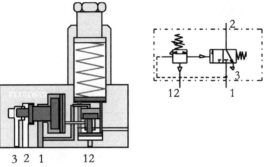

Figure 9.55 | Pressure sequence valve

Example 9.15 | Pneumatic circuit with a pressure sequence valve

A double-acting cylinder is used to press together glued components. Upon pressing a pushbutton, the clamping cylinder is to extend and trip the roller valve. Once the fully extended position of the cylinder has been reached, and sufficient clamping force has been developed, the cylinder is to retract to the initial position. Develop a control circuit using a pressure sequence valve.

The pneumatic circuit to implement the given control task is given in Figure 9.56. The pressure in the sequence valve is set to the working pressure, and the signal input to the pressure sequence valve is tapped from the power line associated with port 4 of the DC valve V1 to the cylinder.

As shown in Figure 9.56(a), valve S1 initiates the forward motion of the cylinder. While the cylinder is moving forward, the pressure in the power line will not build up to the level of the working pressure. Only after the cylinder is fully extended, the maximum clamping pressure develops in the line. When the set pressure in the sequence valve is reached, the integrated 3/2-DC valve is actuated, generating an output signal. The signal is used to reset the final control element V2 and thus causing the return motion of the cylinder, as shown in Figure 9.56(b).

Pneumatic circuit

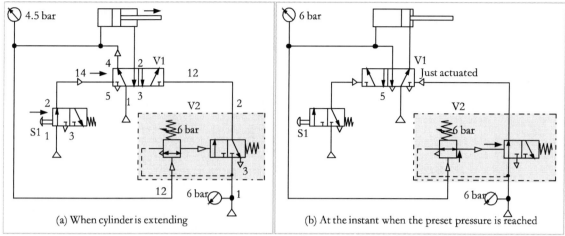

Figure 9.56 | Two positions of the circuit for the control task in Example 9.15

Summary of Controls Pneumatics - Basic Level

Table 9.4 | Summary of controls for Pneumatics - Basic level

Control function	Control valve / Method
Direct control of s/a cylinder or uni-directional motor	3/2 Directional Control (DC) manually-operated valve
Indirect control of s/a cylinder or uni-directional motor	3/2 DC valve, pilot-operated
Speed control of a s/a cylinder	Throttle-check (one-way flow control) valve: • Meter-in method (for Extension) • Meter-out method (for Retraction)
Direct control of d/a cylinder or bi-directional motor	5/2 DC manual valve
Indirect control of d/a cylinder or bi-directional motor without memory	5/2 DC pilot-operated, spring-return valve
Speed control of a d/a cylinder or bi-directional motor	Meter-in method / Meter-out method
Rapid motion of s/a and d/a cylinders	Quick exhaust valve
Memory control of a d/a cylinder	5/2 DC double pilot valve
Semi-automatic operation (Auto-return motion)	3/2 roller-operated valve (Sensor)
Fully-automatic operation with 'Start' and 'Stop' control	3/2 roller-operated valves (Sensors)
Logic operations (OR, AND operations)	Shuttle valve / Two-pressure valve
Time delay between operations	Pneumatic timer
Pressure dependent control	Pressure sequence valve
Counting function	Counter

Objective-type Questions

1. A _____ allows air to flow in one direction and seals it off in the opposite direction.
   a) non-return valve
   b) pneumatic cylinder
   c) adjustable throttle valve
   d) directional control valve

2. Fluid power circuits use schematic drawings to:
   a) make the drawing look impressive
   b) simplify component function details
   c) make an untrained person understand
   d) make it so only trained persons can understand the functions

3. A _____ is used for the direct actuation of a single-acting cylinder.
   a) 2/2 directional control valve
   b) 3/2 directional control valve
   c) 5/2 directional control valve
   d) 5/3 directional control valve

4. A _____ is used to control the direction of the air flow in a single line of a pneumatic control system.
   a) 2/2 directional control valve
   b) 3/2 directional control valve
   c) 5/2 directional control valve
   d) 5/3 directional control valve

5. A _____ is used as the final control element for the control of a double-acting pneumatic cylinder.
   a) non-return valve
   b) 2/2 directional control valve
   c) 3/2 directional control valve
   d) 5/2 directional control valve

6. When this speed control method is employed, the piston of the double-acting pneumatic cylinder is virtually positioned between two air cushions.
   a) by-pass method
   b) meter-in method
   c) bleed-off method
   d) meter-out method

7. Mark the correct statement:
   a) 3/2 DC pneumatic valve has two exhausts
   b) 4/2 DC pneumatic valve has two exhausts
   c) 4/3 DC pneumatic valve has two exhausts
   d) 5/2 DC pneumatic valve has two exhausts

8. What type of logic can be realised by using a shuttle valve?
    a) OR logic
    b) AND logic
    c) NOT logic
    d) Ex-OR logic

9. What type of logic can be realised by using a two-pressure valve?
    a) OR logic
    b) NOT logic
    c) AND logic
    d) Ex-OR logic

10. What is a pneumatic pressure sequence valve?
    a) It is a combination of an adjustable pressure-reducing valve and a check valve
    b) It is a combination of an adjustable pressure-reducing valve and a flow-control valve
    c) It is a combination of the adjustable pressure relief valve and directional control valve
    d) It is a combination of the nonadjustable pressure relief valve and directional control valve

11. What do the numbers in the 5/2 valve mean?
    a) 5 ways and 2 positions
    b) 5 positions and 2 ways
    c) 5 ports and 2 actuating heads
    d) none of the above

Review Questions
1. How is fluid power controlled?
2. What are pneumatic valves?
3. How are pneumatic valves classified?
4. List six possible functions that directional control valves can serve.
5. What is meant by the 'spring-centred' directional control valve?
6. Mention a few features of directional control valves used in signal processing.
7. What are poppet valves?
8. Where are slide valves used?
9. Differentiate between the slide valve and the poppet valve.
10. What are the main features of non-return valves?
11. List out a few derivatives of non-return valves.
12. Explain the operation of a check valve.
13. List three functions performed by pressure-control valves in pneumatic circuits.
14. What is the function of a relief valve?
15. What is the purpose of pressure reducing valve?
16. What are the main types of pressure control valves?
17. Explain the function of pressure sequence valves.
18. Explain the operation of flow-control valves.
19. How is flow control accomplished in flow-control valves?
20. What is a one-way flow-control valve? Explain its function with a neat sketch.
21. In what way a one-way flow-control valve differs from a basic flow-control valve?

22. Draw the ISO symbol of a 5/2-pilot-operated DC valve. Using the numerical system of ISO 5599, designate the ports.
23. Differentiate between 3/2 normally-closed and 3/2 normally-open types of directional control valves.
24. Name four methods of actuating pneumatic directional control valves.
25. Draw as per ISO 1219 the pneumatic symbol for a manually operated 3/2-DC valve (NO type).
26. Where are the three-way valves used?
27. What is the advantage of an internal pilot valve in pneumatic valves?
28. Define the term 'maintained signal' as used in pneumatic systems.
29. Design a pneumatic circuit that operates two double-acting cylinders simultaneously.
30. What is the functional difference between 4/2- and 5/2-way valves, if any?
31. What is meant by 'signal conflict' concerning a 5/2-double-pilot valve? How will it affect the operation of pneumatic circuits?
32. What are the main functions of the sensors?
33. List three factors that determine the speed at which a linear actuator moves.
34. Differentiate between supply-air throttling and exhaust-air throttling.
35. Explain how automatic control is achieved in pneumatic controls.
36. Explain how the speed of the pneumatic cylinder is increased.
37. What is logic control? Give examples of pneumatic logic valves.
38. Draw symbols as per ISO 1219 for valves that give OR function and AND function
39. Explain the function of a two-pressure valve. Give two applications of two-pressure valves.
40. What are the functions of pneumatic time-delay valves? How are time-delay valves classified?
41. Explain the operation of an NC-type time-delay valve. Give an example of it.
42. Explain the function of a shuttle valve. Give an application of a shuttle valve.
43. Explain the structure of a typical pneumatic circuit.
44. What do pressure drops in the fluid power circuit indicate?
Answer key for objective-type questions:
Chapter 9: 1-a, 2-b, 3-b, 4-b, 5-d, 6-d, 7-d, 8-a, 9-c, 10-c, 11-a

9.45 A pneumatically controlled press with a stamping die with a safety guard, as shown in Figure 9.57, is used for producing badges from a very thin metal sheet. A double-acting cylinder is used as the drive for the press. The cylinder is to extend when the safety guard is placed in position, and a pushbutton valve S1 is pressed. The cylinder is to retract automatically after reaching the forward end position and attaining a preset pressure to get consistent quality. The cylinder should retract immediately if the emergency pushbutton valve S2 is pressed. Develop a pneumatic control circuit to implement the control task.

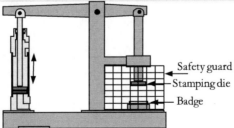

Safety guard
Stamping die
Badge

Figure 9.57 | A stamping die
[Solutions to problems 45, 46, 47, and 48 are given at the end of the book]

9.46 A double-acting cylinder is used for embossing slide rules, as shown in Figure 9.58. The cylinder extends when only two pushbuttons are pressed simultaneously within one second to ensure the complete safety of operators. The cylinder should retract immediately if any one or both pushbuttons are released. Develop a pneumatic circuit to implement the control task.

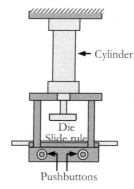

Figure 9.58 | An embossing machine

9.47 Figure 9.59 shows the arrangement for cleaning washers for injection pumps in a cleaning bath. A double-acting cylinder is used to move the container filled with washers up and down in the bath several times. The operator provides a 'Start' signal manually. The washing operation is turned off automatically after a preset time. Develop a pneumatic circuit to implement the control task.

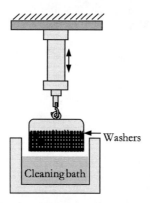

Figure 9.59 | A Cleaning bath for washers

9.48 An arrangement for pressing components for 20 seconds is shown in Figure 9.60. A pushbutton PB1 is used to control the forward stroke of the cylinder. After pressing the components for 20 seconds, the cylinder is to retract automatically. The return stroke must occur even if the start pushbutton is still depressed. A new start signal may only be effective after the initial position of the piston is reached, and the pushbutton is released. Develop a pneumatic circuit to implement the control task.

Figure 9.60 | An arrangement for pressing components

Chapter 10 | Pneumatic Applications

The areas of pneumatic applications are quite vast. Pneumatic systems and devices are increasingly being used in all modern manufacturing and processing industries to perform a wide variety of work operations. Pneumatics finds application in industry sectors ranging from automotive manufacturing to onboard commercial vehicles, from rail applications to printing and textiles, from food packaging to process industries, from the electronic sector to medical care, and in thousands of other specialised industries.

Why Pneumatics?

It is practically impossible to move around without seeing some applications of pneumatics. The widespread use of pneumatics has been made possible due to the easy availability of standard components, such as cylinders, rotary drives, and valves, providing solutions for many applications, thus requiring only less time for planning, assembling, commissioning, and maintenance. Pneumatic cylinders and valves are modular elements, which can easily be integrated into systems to provide solutions for many control tasks. Pneumatic systems are quite amenable to automation with an affordable degree of flexibility. Therefore, pneumatic drives can cover a large area of applications. The ability of pneumatic elements to do useful work economically and efficiently is responsible for their widespread use.

Pneumatic Applications

Machines, equipment, and tools powered by compressed air are adaptable for operations requiring a direct application of force on the work-piece such as in clamping, riveting, embossing, forming, drawing, cutting, polishing, etc. Pneumatic holding devices such as power chucks, collets, and mandrels are widely used for machining operations where work-piece must be held securely and accurately. Pneumatic portable power tools perform a wide range of activities such as nut running, screw driving, grinding, drilling, etc.

With the growing demand for automatically controlled plants and machinery, the potentialities of pneumatics for power as well as control applications are increasingly being realised and put into practice. Typical applications are with handling functions where work-pieces are transferred from one section to the next section in completely automated facilities. Air springs can be used as good shock absorbers as these absorb more energy than mechanical devices of comparable size. They can be adjusted to varying conditions of velocity and inertia.

There are numerous applications where a vacuum is used to perform useful functions. Industrial applications where vacuum devices are used include materials handling, clamping, and forming. In materials handling applications, Grippers, vacuum devices, and suction cups are used to lift lightweight work-pieces, like glass plates, sheet metal, sheets of paper, ceramic tiles, etc., which have flat and clean surfaces. For example, spring-loaded vacuum suction cups can be used to lift plywood and hard fibreboard panels from stacks and place these on a conveyor belt.

Pneumatically-powered tools, machines, and systems are found in the industrial and automobile sectors.

- **Industrial Applications:** Production systems are undergoing a significant change with products becoming more complex, customers demanding instant delivery, and product life cycles becoming shorter. Production systems for material handling, injection moulding, packaging, and metal forming systems use standardised pneumatic components for drilling, sawing, clamping, feeding, lifting, pressing, positioning, bending, turning, gripping, ejecting, conveying, assembling, and testing operations.
- **Automobile:** The automobile industry uses pneumatic systems for dismantling vehicle tires, vehicle painting, opening and closing doors, air brakes on heavy vehicles, etc.

Objective-type Questions

1. Mark the <u>correct</u> statement:
   a) Pneumatic systems are difficult to automate
   b) Pneumatic portable power tools can be used for nut running, screw driving, grinding, and drilling
   c) Most of the components for pneumatic systems are custom-made as it is difficult to get standard components
   d) Pneumatic systems are used in industrial applications as they can provide huge forces to the tune of 100 tonnes, most efficiently

Review Questions

1. What are the reasons for the widespread use of pneumatics in modern manufacturing and processing industries?
2. Explain with the help of a neat sketch, the method of amplifying the force (thrust) of a given pneumatic cylinder.
3. What are the various methods to increase the stroke of a given cylinder?
4. Briefly explain the issues involved in the selection of the following for a given application: (1) Pneumatic cylinder (2) Pneumatic valve.
5. Give examples for pneumatic actuators used for the following types of motion:
   - Linear motion
   - Angular motion
   - Continuous rotary motion
   - Intermittent rotary motion
6. Classify various industrial work operations. Why is pneumatics best suited for these operations?
7. Name three applications of vacuum devices.
8. What are the essential factors to be taken into account while designing a pneumatic system?
9. What objectionable outcome takes place when components of pneumatic systems are undersized?
10. What unwanted result happens when components of pneumatic systems are oversized?
11. Name a few important considerations that must be taken while selecting pneumatic actuators.
12. Name a few important considerations that must be taken while selecting pneumatic power valves.
13. Explain the terms: (1) Anti-repeat and (2) anti-tie down concerning the designing of a machine' controls

Answer key for the objective-type question:
Chapter 10: 1-b

Chapter 11 | Maintenance, Troubleshooting, and Safety Features of Pneumatic Systems

In general, the term 'maintenance' of a system covers a broad range of routine maintenance and repair activities, intended to keep the system in a satisfactory working condition. The same broad definition applies to the specialised area of pneumatic machines and systems as well.

Classification of Maintenance

Maintenance can be classified into two basic categories:

- Preventive (or proactive) maintenance
- Corrective (or breakdown or reactive) maintenance

Preventive maintenance is undertaken when a machine is operating correctly to prevent any potential failure of the machine. It is performed regularly, as per a schedule or a checklist. This action ensures the efficient working of all components of the machine, at all times. The objective of performing the preventive maintenance of a machine is to prolong its useful service life.

On the other hand, corrective maintenance is undertaken on a machine after its failure. This activity consists of finding the fault and repairing the machine. It is needless to say that preventive maintenance is the most effective maintenance strategy, and one has to focus on the prevention of failure rather than troubleshooting the machine. It is also essential to take adequate safeguards to prevent personal injury and damage to pneumatic machines during maintenance and troubleshooting activities.

Definitions of Maintenance Activities

In general, maintenance involves certain closely-related activities, such as inspection, servicing, examination, and overhaul.

- Inspection refers to the maintenance activity that comprises a careful visual observation/scrutiny of the machine, usually without dismantling it.
- Servicing refers to the cleaning, adjustment, lubrication, and other servicing functions of the machine without dismantling it.
- Examination relates to the inspection of the machine with necessary dismantling, measurement, and non-destructive tests to obtain useful information regarding the condition of the components/subassemblies of the machine.
- Overhaul refers to the extensive work done to repair and/or replace the worn-out and defective parts of the machine.

Requirements for Preventive Maintenance

The most general requirements for preventive maintenance to prevent failure or breakdown are as follows:

- Know the machine
- Understand and follow the best maintenance practices
- Compile a maintenance checklist
- Follow instruction manuals
- Ensure safety
- Stock spares

Malfunctions in Pneumatic Systems

Pneumatic systems are neither perfect nor immune to failures. Faults in pneumatic systems can be attributed to many reasons. Faults in a pneumatic system are generally due to stoppage/slower performance (lack of force), poor performance (low speed), erratic operation, and/or leakages. Another potential reason for the faults is the presence of dust, moisture, etc. Each conductor connection should be periodically checked for its tightness. A general account of malfunctions is given below:

Malfunctions due to Contaminants

Rust and scale particles are introduced within the piping of a pneumatic system by moisture resulting from the condensation of moisture. The free-moving particles combined with oil and water sludge can scratch seals and abrade surfaces of precision-made parts of valves and cylinders in the system, thus causing leaks. The particles can also block orifices and cause valve spools to jam. Further, flow passages may become restricted, resulting in a reduced flow rate and increased pressure drop. The moisture present can also wash away lubricants from the valves, resulting in the faulty operation of the valves, corrosive damage to the valve surfaces, and excessive wear of system components.

Malfunctions due to Under-sized Air Supply

Many a time, pneumatic systems/machines in a factory or workshop are added without enlarging the capacity of the existing compressed air supply, or the compressed air pipe size is too small. Due to this, malfunctions can occur sporadically.

Malfunctions due to Under-lubrication/Over-lubrication

The absence of lubrication or under-lubrication in a system will cause increased wear and the consequent deterioration of the system components. Over lubrication may produce the sluggish operation of valves, cylinders, and pneumatic tools.

Malfunctions due to Improper Mountings

A cylinder mounted incorrectly will produce an undue strain on its mounting plate and mounting bolts. The piston-rod of a cylinder that is not adequately supported or not correctly aligned with the centreline of the associated load will exert severe strain on its seals and glands.

Faults in Pneumatic Systems

The more specific reasons for these faults in pneumatic systems are:
- Misalignment or mechanical jam
- Power supply failure
- Insufficient pressure or low voltage,
- Twisted tubing
- Burned solenoid coils
- Failure of arc suppression circuits
- Bend piston-rods/barrel
- Flow restrictions
- Lack of lubrication
- Insufficient compressed air deliver

The erratic operation can arise from sticking valves or due to any mismatch in the total requirement of compressed air by the system and the actual compressor delivery volume.

Consequences of Poor Maintenance

Lack of regular maintenance may result in the loss of air and associated pressure drops, premature wear of moving parts, production shortfalls, and increased downtime of pneumatic components. These effects eventually result in increased downtime of the system and shortfalls in production.

Preventive Maintenance of Pneumatic Systems

The preventive maintenance of a system is to be carried out by taking into account the individual components of the system.

Maintenance of Compressors

The maintenance of compressors must be carried out following the manufacturer's instructions. A compressor should be located in a clean, accessible area for its visual inspection and maintenance. The essential routine maintenance activities of a compressor are cleaning, visual inspection, running checks, and servicing of filters, lubricators, and coolers. Regular checking of the lube oil level in a compressor is an essential maintenance task. A compressor with an integrated coolant needs to be regularly checked for the inlet and outlet temperatures of the coolant. Many compressors are belt-driven and require belt conditions to be checked at regular intervals.

Maintenance of Air Receivers

Condensate drains from air receivers should be automatic where possible, but they still need to be inspected regularly to make sure that the complete unit is working correctly. Safety devices on air receivers like pressure relief valves must be maintained in satisfactory functional order. All special pressure vessel rules about air receivers must be observed fully.

Maintenance of Air-Mains

Proper maintenance of air mains is essential for the effective removal of contaminants. An essential requirement of any air mains network must be to stop leakage, as far as possible.

Regular inspection of air mains for leaks when there is no background noise. A method of detecting the leakage is applying soapy water or commercially available leak-detecting liquids like aerosol sprays on suspected joints. Another method is to use an ultrasonic leak detection instrument. When air leaks, it moves from a high-pressure side to a low-pressure side through the leak spot where it expands quickly and creates a turbulent flow. This turbulence has strong ultrasonic components. The intensity of the ultrasonic signals falls off quickly from the source, permitting the exact spot of the leak to be detected.

Maintenance of Air Service Unit (FRL Unit)

If an air service unit (FRL unit), consisting of a filter, regulator, and lubricator, is not maintained correctly, the investment made on the unit and its installation, turn out to be a mere waste. The following regular maintenance of FRL is of utmost importance:

Filter

The condensate level in the filter must be checked regularly. It must not exceed the maximum level marked. If the condensate level exceeds the maximum level, the condensate is liable to be drawn into the air stream again. Therefore, the accumulated condensate must be drained, before reaching the maximum level, either automatically or manually by opening the drain screw. The filter element should be removed and replaced when the filter is clogged.

Regulator
Usually, this unit requires no regular maintenance, especially during the initial years of its service life.

Lubricator
The oil in the lubricator, if used, is consumed in the process of lubricating the compressed air. Check the oil level in the lubricator, and top up, if necessary.

Maintenance of Pneumatic Cylinders
Apart from the general maintenance activities, the following maintenance activities can be carried out on a cylinder to keep it in good working condition:

- Check the piston-rod for straightness. Check for any dents or damages on the piston-rod due to impact forces,
- Examine the piston-rod bearing for roundness,
- Examine the barrel, the piston, and the piston-rod for nicks, scoring, and pitting,
- Check the cylinder for worn components,
- Check and control the leakages in the cylinder,
- Replace piston seals, piston-rod seals, and piston-rod bearings, if leakages occur,
- Align the cylinder and its mating part in line, to avoid side loads on the cylinder,
- Check the cylinder mountings periodically for tightness or cracks,
- Check for sluggish/erratic operation of the cylinder,
- Check for the creeping of the cylinder, and
- Replace the spring in the single-acting cylinder, if broken.

Maintenance of Pneumatic Valves
As pneumatic valves are manufactured with tight-fitting delicate parts, contaminants can pose significant problems for the valves. Small amounts of dirt, rust, and sludge can lodge in between the mating surfaces of the valves causing their abrasion, seal damage, and internal leakage. The contaminants are also responsible for the sticking of valve spools and the plugging of small openings in valves.

Another probable causes of valve failure are jammed springs. Valves can also be damaged, in particular, their operating mechanisms, by incorrect installation and operation. Next, the coils of solenoid valves are affected by vibration, moisture, and corrosion.

Troubleshooting Pneumatics
Preventive maintenance is carried out on pneumatic systems to keep them in perfect working condition at all times. However, faults/breakdowns do occur in pneumatic systems, which have to be traced and corrected with minimum delay and expense. In general, pneumatic failures can be attributed to the presence of contaminants, the clogging of filters, and loose connections. The symptoms of the failures are generally manifested in the form of the development of excessive pressure drops, heat, and noise. A useful troubleshooting strategy needs to be in place for the quick detection and rectification of faults in pneumatic systems.

General Troubleshooting Procedure
In general, the fault-finding and repair of a faulty system encompass many activities to minimising the time and cost involved in the fault-finding and repair. These activities include collecting information on

the fault, analysing and evaluating the information, localising the fault, conducting necessary tests, and repairing the fault.

The first step in troubleshooting a system that has developed some trouble is to understand its operation and associated circuits. Once sufficient information is collected and evaluated, visualise all possible root causes of the fault.

The most straightforward test may be conducted for finding the section that contains the defective component of the circuit. A careful check of the components involved in this part may lead to the source of trouble. If not, the cycle is repeated until the fault is traced and repaired.

The final steps of troubleshooting are concerned with fault recording and fault analysis to discover any recurring pattern of faults or any design and application problem or any shortcomings in the relevant knowledge of maintenance personnel if any.

Safety in Pneumatic Systems

Nowadays, virtually every industrial production machinery or process equipment employs pneumatic components. A considerable amount of energy contained in compressed air, high speed and power of pneumatic actuators, and unexpected loss of pressure pose a danger to the equipment and personnel. Observe all safety procedures when working with pneumatic systems and implement necessary legal requirements for the system.

Air receivers used downstream of compressors can store a tremendous amount of energy. The main reason for many of the previous air receiver explosions appeared to be the result of the ignition of oil carried over from the compressor. The explosion is only likely to occur if the compressed air gets hot enough to cause spontaneous combustion.

The energy stored in a long tube or pipe will be expelled through its open end in a very short time with enough velocity and force, to cause a severe whiplash of the line.

Protection against a loss of pressure on such items as chucks and vices is usually provided for by making them self-locking so that they will be released only if a force is applied in the reverse direction.

One of the most hazardous types of operation is where the operator's hands have to be inserted into the working area to feed or eject a component. The control system for pneumatic clamping devices should be designed and arranged in such a way as to avoid accidental operation.

There is a possibility of danger to personnel and equipment if a machine under load stops due to the failure of the air supply. Fail-safe circuits can be designed to protect the operator and the machine against electrical power supply failure, air supply failure, overload, carelessness, etc. Generally, they are designed to return the load to some 'safe' condition should one of these unscheduled events occur. Circuits should be designed, in general, to prevent the inadvertent operation of pneumatic devices.

In any fault tracing exercise, the personal safety and safety of others are paramount. Work should be carried out using approved practices and observing the relevant legislation. Any moving parts should be mechanically locked, and trapped air in any section should be exhausted.

Objective-type Questions

1. The general requirements of good preventive maintenance are:
   a) Operational skills and time management
   b) Following instruction manuals and carrying out maintenance as per the checklist
   c) System design capability and ability to follow best practices
   d) Knowledge of machine operation, analytical ability, and ensuring safety

2. Faults in a pneumatic system are due to:
   a) Lack of force or speed
   b) Erratic operation
   c) Leakages
   d) All of the above

3. Mark the <u>incorrect</u> statement:
   a) A pneumatic system can malfunction due to the undersized air supply, contaminants, lack of lubrication, and loose connections
   b) A compressor with an integrated reservoir and aftercooler is a maintenance-free unit.
   c) Pneumatic cylinder maintenance involves checking parts for dents, nicks, scoring, pitting, etc., replacing damaged sealing, busing, etc., and tightening the loose mounting
   d) An essential troubleshooting procedure for pneumatic systems involves gathering information about the system, finding the section of the system that contains the defective part, and repairing the defective component.

Review Questions

1. What is maintenance? How is it classified?
2. What is the importance of preventive maintenance, and what are its advantages?
3. What is the most significant cause of pneumatic system failure?
4. What is the importance of instruction manuals supplied by manufacturers for maintenance personnel?
5. What are the factors upon which the intervals of maintenance activities depend?
6. How necessary is the lubrication of pneumatic components?
7. What are the usual causes of the failure of pneumatic equipment?
8. Explain briefly the malfunctions caused due to (a) contaminants, (b) improper mountings, (c) over-lubrication, and (d) under-lubrication.
9. Write a brief note on the maintenance of the following: (a) Compressors, (b) Air cylinders, and (c) Valves.
10. Why is it essential to carry out regular maintenance on FRL?
11. Write down a few maintenance activities that are carried out on pneumatic cylinders.
12. Write down a few maintenance activities that are carried out on pneumatic valves.
13. What is the procedure adopted for troubleshooting a circuit that has been in existence for a long time?
14. Explain the importance of a good troubleshooting strategy.
15. What is the meaning of 'two-hand safety operation' with regard to hazardous types of operation?

Answer key for the objective-type questions:
Chapter 11: 1-b, 2-d, 3b

Appendix 1

Conversion Tables

Conversion factors for various physical terms may be taken from these tables.

Length

	cm	metre	km	in	ft
1 centimetre	1	0.01	10-5	0.3937	0.03281
1 metre	100	1	0.001	39.37	3.281
1 kilometre	100000	1000	1	39370	3281
1 inch	2.54	0.0254	2.54x10-5	1	0.08333
1 foot	30.48	0.3048	0.0003048	12	1

Area

	sq metre	sq cm	sq ft	sq in
1 square metre	1	10000	10.76	1550
1 square centimetre	0.0001	1	0.001076	0.1550
1 square foot	0.0929	929.0304	1	144
1 square inch	0.0006452	6.4516	0.006944	1

Volume

	cu metre	cu cm	l	cu ft	cu in
1 cubic metre	1	1000000	1000	35.31	61023.74
1 cubic centimetre	10-6	1	0.001	3.531x10-5	0.06102
1 litre	0.001	1000	1	0.03531	61.02374
1 cubic foot	0.02832	28316.85	28.31685	1	1728
1 cubic inch	1.639x10-5	16.38706	0.01638706	0.0005787	1

Mass

	g	kilogram	pound
1 gram	1	0.001	0.002204623
1 kilogram	1000	1	2.204623
1 pound	453.5924	0.4535924	1

1 metric ton = 1000 kg.

Force

	dyne	Newton	lb	kgf
1 dyne	1	10-5	2.248x10-6	1.019716x10-6
1 Newton	100000	1	0.2248089	0.1020
1 pound	444822.2	4.448222	1	0.4535924
1 kilogram-force	980665	9.80665	2.204623	1

Pressure

	Pascal	bar	kgf/cm²	atmosphere	psi
1 Pascal	1	10-5	1.02x10-5	9.87x10-6	0.000145
1 bar	100000	1	1.02	0.987	14.50377
1 kgf/cm²	98066.5	0.980665	1	0.9678411	14.22334
1 atmosphere	101325	1.01325	1.033227	1	14.69595
1 pound/in² (psi)	6894.757	0.06895	0.070307	0.06804596	1

1 torr = 1 mm of Hg.

Volume Flow Rate

	m³/s	m³/h	ft³/m	ft³/h	lit/s	lit/h
1 m³/s	1	3600	2118.88	127132.8	1000	3600000
1 m³/h	0.0002778	1	0.58858	35.31467	0.2777778	1000
1 ft³/m	0.0004719	1.699011	1	60	0.4719474	1699.011
1 ft³/h	7.8658x10-6	0.028317	0.016667	1	0.0078658	28.31685
1 lit/s	0.001	3.6	2.11888	127.1328	1	3600
1 lit/h	2.7778x10-7	0.001	0.000589	0.035315	0.00027778	1

Temperature

$$°C = (°F - 32) \times 5/9$$
$$°F = °C \times 9/5 + 32$$

Multiples

Multiples	
deca	10
hecto	10^2
kilo	10^3
mega	10^6
giga	10^9
tera	10^{12}

Submultiples	
deci	10^{-1}
centi	10^{-2}
milli	10^{-3}
micro	10^{-6}
nano	10^{-9}
pico	10^{-12}

Appendix 2

Typical Compressor Data

A2.1 | Single-stage Compressors (Max Pressure 9 bar)

Table A2.1 | Single-Stage compressors (Max Pressure 9 bar)

Power	Drive speed	Displacement	FAD @6bar	Air Receiver
KW	rpm	lpm	lpm	litres
2.2	550	360	256	160
3.7	925	605	443	220
5.5	690	921	708	220
7.5	920	1228	950	220
7.5	920	1228	950	420
11	925	1853	1390	500
15	925	2406	1925	450
18.6	1050	3078	2463	450

A2.2 | Two-stage Compressors (Max Pressure 12 bar)

Table A2.2 | Two-Stage compressors (Max Pressure 12 bar)

Power	Drive speed	Displacement	FAD @6bar	Air Receiver
KW	rpm	lpm	lpm	litres
2.2	925	311	250	160
2.2	925	311	250	220
3.7	925	501	410	220
5.5	1050	700	580	220
7.5	750	997	850	220
7.5	750	997	850	420
7.5	750	997	850	500
11	1150	1535	1250	500
15	1150	2195	1756	450
18.6	1150	2973	2379	450
22.4	1150	3363	2690	450

A rule of thumb suggests that on a steady pumping, a compressor will produce a minimum of 113 Nlpm flow of air for every HP capacity at 6 bar.

Appendix 3

A3.1 | Thrusts and Pulls of Single-acting Cylinders

Table A3.1 | Thrusts and pulls of single-acting cylinders

Cylinder bore, mm	The minimum pull of spring, N	Thrust, N at 6 bar
10	3	37
12	4	59
16	7	105
20	14	165
25	23	258
32	27	438
40	39	699
50	48	1102
63	67	1760
80	86	2892
100	99	4583

A3.2 | Thrusts and Pulls of Double-acting Cylinders

Table A3.2 | Thrusts and pulls of double-acting cylinders

Cylinder bore mm (inches)	Piston rod dia mm (inches)	Thrust, N (at 6 bar)	Pull, N (at 6 bar)
8	3	30	25
10	4	47	39
12	6	67	50
16	6	120	103
20	8	188	158
25	10	294	246
32	12	482	414
40	16	753	633
44.43 (1¾)	16	931	810
50	20	1178	989
63	20	1870	1681
76.2 (3)	25	2736	2441
80	25	3015	2721
100	25	4712	4418
125	32	7363	6881
152.4 (6)	(1½)	10944	10260
160	40	12063	11309
200	40	18849	18095
250	50	29452	28274
304.8 (12)	(2¼)	43779	42240
320	63	48254	46384
355.6 (15)	(2¼)	59588	58049

Note: For pressures other than 6 bar, multiply the thrust/pull at 6 bar by the given absolute pressure and divide it by 7. These figures do not account for the seal or packing friction in these cylinders. This type of friction is estimated to affect the thrust of the cylinders by about 10%.

Table A3.3 | Air consumption of pneumatic cylinder

Bore mm	Rod mm	Air consumption for the		
		forward stroke of 1 mm at 6 bar dm³/mm	return stroke of 1 mm at 6 bar dm³/mm	combined strokes of 1 mm at 6 bar dm³/mm
10	4	0.00054	0.00046	0.00100
12	6	0.00079	0.00065	0.00144
16	6	0.00141	0.00121	0.00262
20	8	0.00220	0.00185	0.00405
25	10	0.00344	0.00289	0.00633
32	12	0.00563	0.00484	0.01047
40	16	0.00880	0.00739	0.01619
50	20	0.01374	0.01155	0.02529
63	20	0.02182	0.01962	0.04144
80	25	0.03519	0.03175	0.06694
100	25	0,05498	0.05154	0.10652
125	32	0.08590	0.08027	0.16617
160	40	0.14074	0.13195	0.27269
200	40	0.21991	0.21112	0.43103
250	50	0.34361	0.32987	0.67348

Note:
1. *Take each figure and multiply by the stroke in mm.*
2. *For pressures other than 6 bar, multiply the air consumption value by the given absolute pressure and divide it by 7.*

To estimate the total average air consumption of a typical pneumatic system, calculate the air consumption for each cylinder in the system using the formulae given above. Add the estimated air consumption of all cylinders and add 5% to make allowance for the loss due to leakage and friction.

A3.4 | Essential Specifications of Actuators

The essential technical specifications for pneumatic actuators are given in Table A3.4.

Table A3.4 | Typical specifications for pneumatic components:

1.	Medium	e.g., Compressed air, filtered, lubricated
2.	Operation	e.g., doubling-acting, air-cushioned
3.	Operating pressure	e.g., 0.1 to 12 bar
4.	Operating temperature	e.g., -10°C to +80°C
5.	Linear actuators:	
5.1	Size	Piston rod diameters from 1 mm to 320 mm
5.2	Thrust	2.7 N to 48000 N (at 6 bar)
5.3	Stroke length	1 mm to 10 m
5.4	Speed	5 to 15000 mm/s
6.	Rotary actuator:	
6.1	Size	Rotary drive diameter from 6 mm to 100 mm
6.2	Torque	0.15 Nm to 150 Nm (at 6 bar)
6.3.	Angle of rotation	1° to 360°
6.4.	Speed	Up to 50000 rpm

Appendix 4

Graphic Symbols for Pneumatic Components

A4.1 Supply Elements

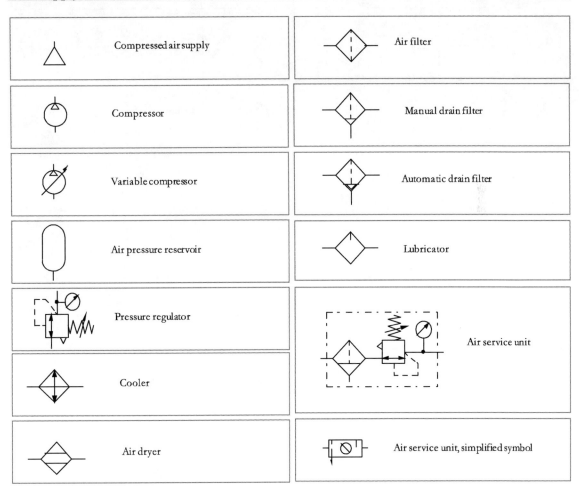

Compressed air supply	Air filter
Compressor	Manual drain filter
Variable compressor	Automatic drain filter
Air pressure reservoir	Lubricator
Pressure regulator	Air service unit
Cooler	
Air dryer	Air service unit, simplified symbol

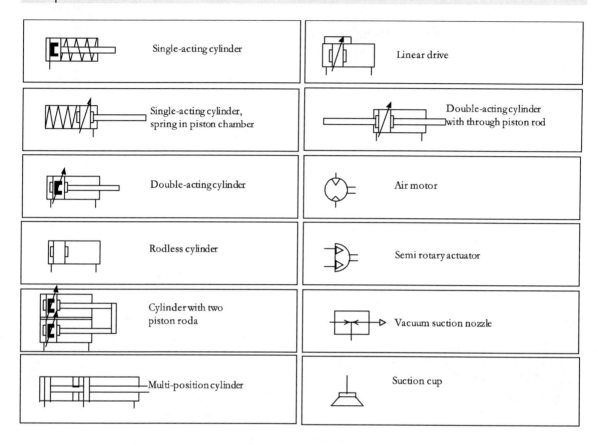

Single-acting cylinder	Linear drive
Single-acting cylinder, spring in piston chamber	Double-acting cylinder with through piston rod
Double-acting cylinder	Air motor
Rodless cylinder	Semi rotary actuator
Cylinder with two piston roda	Vacuum suction nozzle
Multi-position cylinder	Suction cup

	2/2-way valve (NC) type			Check valve
	2/2-way valve (NO) type			Spring-loaded check valve
	3/2-way valve (NC) type			Orifice valve, adjustable type
	3/2-way valve (NO) type			Shuttle valve
	4/2-way valve			Two-pressure valve
	5/2-way valve			Quick-exhaust valve
	5/3-way valve, all closed centre position			Throttle valve with constant restriction
	5/3-way valve, open exhaust centre position			Throttle valve with adjustable flow control
	5/3-way valve, open pressure centre position			Throttle check valve, adjustable

Working line		Energy tapping point closed	
Control line		Pressure gauge	
Flexible line		Pressure indicator	
Line connection, rigid		Flow meter	
Line crossover		Quick release coupling uncoupled without check valve	
Exhaust without pipe connection		Quick release coupling connected without check valve	
Exhaust with pipe connection		Quick release coupling uncoupled with check valve	
Silencer		Quick release coupling connected with check valve	

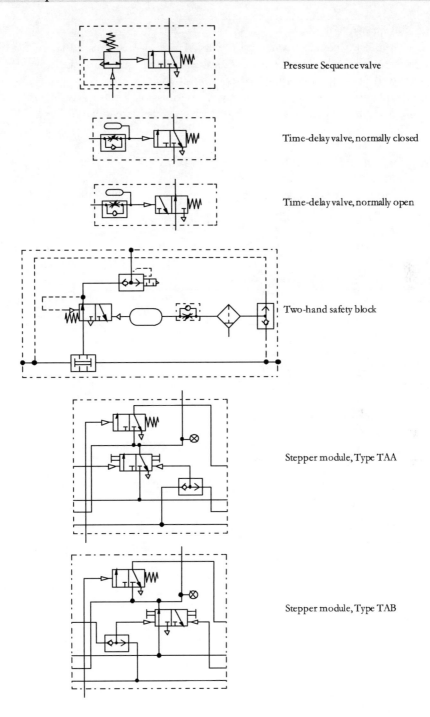

Pressure Sequence valve

Time-delay valve, normally closed

Time-delay valve, normally open

Two-hand safety block

Stepper module, Type TAA

Stepper module, Type TAB

Appendix 5

Air Quality Classification

ISO 8573-1: 2010 stipulates contaminants and quality classes of compressed air for general use. Air contains solid, water, and oil particles as contaminants. The standard specifies the amount of contamination allowed in each cubic metre of compressed air. A quality (or purity) class number is defined for each of these contaminants according to the permissible levels of specific parameters. These parameters and their permissible values of them against each class are given in Tables A5.1, A5.2, and A5.3.

A. Solid Particles

Table A5.1 | Permissible levels of solid contaminants (Ref: ISO 8573-1)

Class (A)	Max. particles per m³			Remarks
	0.1- 0.5 µ*	0.5-1 µ*	1-5 µ*	
1	0 - 20000	0 - 400	0 - 10	Highest quality
2	20001 - 400000	401 - 6000	11 - 100	Better than class 3
3	-	6001 - 90000	101 - 1000	Better than class 4
4	-	-	1001 - 10000	Better than class 5
5	-	-	10001 - 100000	Better than class 6
6	With the mass concentration of >5µ particles $\leq$5 mg/m³			Better than class 7
7	With the mass concentration of >5µ particles 5-10 mg/m³			Better than class x
x	With the mass concentration of >5µ particles >10 mg/m³			Lowest quality

*Size range of solid particles

Note: Class is assigned for the worst-case situation.

B. Water

Table A5.2 | Permissible levels of water contaminant (Ref: ISO 8573-1)

Class (B)	Vapour pressure dew point	Liquid water g/m³	Remarks
1	$\leq$-70°C	-	Use class 1 desiccant dryer
2	$\leq$-40°C	-	Use class 2 desiccant dryer
3	$\leq$-20°C	-	Use class 3 desiccant dryer
4	$\leq$+3°C	-	Use class 4 refrigerant dryer
5	$\leq$+7°C	-	Use class 5 refrigerant dryer
6	$\leq$10°C	-	Use class 6 refrigerant dryer
7	-	$\leq$0.5	Better than class 8
8	-	0.5-5	Better than class 9
9	-	5-10	Better than class x
x	-	>10	Lowest quality

C. Oil Particles

Table A5.3 | Permissible levels of oil particles (Ref: ISO 8573-1)

Class (C)	mg/m³	Remarks
1	0.01	Highest quality
2	0.1	Better than class 3
3	1	Better than class 4
4	5	Better than class x
x	>10	Lowest quality

Class 0

A quality class 0 can be defined as per a written specification between the user and the supplier. It should be more stringent than class 1.

Explanation

An air quality class is specified as a combination of the three air quality numbers in the format: A.B.C.

For example, a quality class 1.2.1 means that in each m³ of compressed air, the particulate count should not exceed 20000 particles in the 0.1-0.5 μ size range, 400 particles in the 0.5-1 μ size range and 10 particles in the 1-5 μ size range.

A pressure dewpoint of -40°C or better is required, and no liquid water is allowed. In each cubic metre of compressed air, not more than 0.01mg of oil is allowed.

This weight of the oil is the total level for liquid oil, oil aerosol, and oil vapour.

Solutions to Selected Numerical Problems

Chapter 2: Question 11

A mass of 500 Kg needs to be pushed upwards using a double-acting pneumatic cylinder. What diameter of cylinder do we need if the pressure available is 6 bar?

Solution
Clamping force = 500 x 9.87 = 4935 N
Operating pressure, P = 6 bar = 6 x 10^{-5} Pa

Area of the cylinder, A = F/P = 4935 / 6 x 10^{-5} m^2 = 822.5 x 10^5 = 0.008225 m^2

Bore diameter, cylinder, D = $\sqrt{\frac{4 \times A}{\pi}}$ = $\sqrt{\frac{4 \times 0.008225}{\pi}}$ = 0.1 m = 100 mm

Chapter 2: Question 12

Determine the bore size of a pneumatic cylinder to keep a work-piece pressed with a clamping force (F) of 1000 N statically during the extension stroke and at the operating pressure (P) of 5 bar. Consider the load factor as 0.7. Note: The actual output force produced by a cylinder is lower than the theoretical force output (F=PxA) by its load factor due to its frictional and sliding resistance.

Solution
Clamping force = 1000 N
Load factor = 0.7
Theoretical Force to be developed by the cylinder, F = 1000/0.7 = 1429 N
Operating pressure, P = 5 bar = 5 x 10^{-5} Pa

Area of the cylinder, A = F/P = 1429 / 5 x 10^{-5} m^2 = 285.8 x 10^5 = 0.002858 m^2

Bore diameter, cylinder, D = $\sqrt{\frac{4 \times A}{\pi}}$ = $\sqrt{\frac{4 \times 0.002858}{\pi}}$ = 0.06 m = 60 mm

Chapter 4: Question 28

An air reservoir of 3 cubic metres is connected to a compressor, which delivers 6 cubic metres of air per minute. What will be the pressure in the tank, measured by a pressure gauge after three minutes?

Solution
Size of the reservoir   = 3 m^3
Delivery rate of the compressor   = 6 m^3/min
Original volume of air in the reservoir after 3 minutes, V1 = 3+6+6+6 = 21 m^3
Original pressure, P1   = 1 bar(a)

Final volume of air in the reservoir, V2   = 3 m^3
Pressure in the reservoir after 3 minutes, P2 = P1 x V1 /V2 = 1 x 21 / 3
                              = 7 bar(a) = 6 bar(g)

Chapter 8: Question 7

Determine the theoretical thrust of a 320 mm bore double-acting pneumatic cylinder supplied with compressed air at a pressure of 7 bar.

Bore diameter = 320 mm
Pressure = 7 bar

Thrust, $F = \dfrac{\pi \cdot 320^2}{40} \cdot 7 = 56297$ Newton

Chapter 8: Question 8

Calculate the air consumption (V) of a double-acting cylinder, with a 50 mm bore, 20 mm piston-rod diameter, and stroke length of 500 mm. The cylinder is supplied with compressed air at a pressure of 6 bar.

Bore diameter, D = 50 mm
Rod diameter, d = 20 mm
Stroke length, S = 500 mm
Pressure, Ps = 6 bar
Atmospheric pressure = 1 bar

$$V(\text{out-stroke}) \text{ in litre} = \frac{\pi D^2}{4} \; S \; \frac{Ps+Pa}{Pa} \; 10^{-6}$$

$$V(\text{out-stroke}) \text{ in litre} = \frac{\pi \times 50^2}{4} \; 500 \; \frac{6+1}{1} \; 10^{-6}$$

=6.87 litre

$$V(\text{in-stroke}) \text{ in litre} = \frac{\pi(D^2 - d^2)}{4} \; S \; \frac{Ps+Pa}{Pa} \; 10^{-6}$$

$$V(\text{in-stroke}) \text{ in litre} = \frac{\pi(50^2 - 20^2)}{4} \; 500 \; \frac{6+1}{1} \; 10^{-6}$$

=5.77 litre

Air consumption for double stroke, per cycle = V(out-stroke) + V(in-stroke) = 6.87 + 5.77 =12.64 litre

Solutions to Problems 45. 46. 47, and 48 of Chapter 9

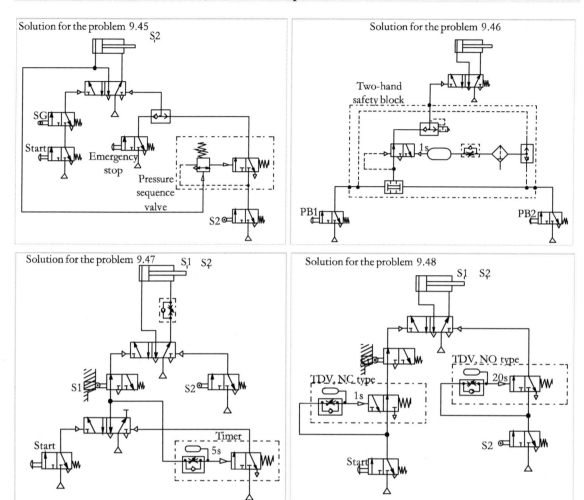

Solution for the problem 9.45
S2
SG
Start
Emergency stop
Pressure sequence valve
S2

Solution for the problem 9.46
Two-hand safety block
1 s
PB1
PB2

Solution for the problem 9.47
S1 S2
S1
S2
Start
Timer
5s

Solution for the problem 9.48
S1 S2
TDV, NC type
1s
TDV, NO type
20s
S2
Start

12 | References

1. Air Compressor Guide - Getting the Most for Your Money, How to Select and Protect Your Air Compressor Investment, Kaeser Compressors, Inc. P.O. Box 946, Fredericksburg, VA 22404, www.kaeser.com

2. Andrew Parr, Hydraulics & Pneumatics, A technician's and Engineer's Guide, 2nd Edition, Butterworth, Heinemann, 1998

3. Anthony Esposito, Fluid Power with Applications, 6th Edition, Prentice-Hall of India, 2006

4. Compressed Air Engineering, Basic principles, tips and suggestions, KAESER KOMPRESSOREN SE, P.O. Box 2143 – 96410, Coburg, GERMANY, www.kaeser.com

5. Compressed Air System Guide, Designing Your Compressed Air System, How to Determine the System You Need, Kaeser Compressors, Inc. P.O. Box 946, Fredericksburg, VA 22404, www.kaeser.com

6. Compressed Air System Installation Guide, Layout Considerations for a Reliable, Energy Efficient, and Safe Compressed Air System, by Kaeser's Compressed Air and Engineering Experts, Kaeser Compressors, Inc.511 Sigma Drive, Fredericksburg, Virginia 22408 USA, www.us.kaeser.com

7. Compressed Air Treatment Guide - Meeting Your Compressed Air Treatment Needs, How to Select the Right Equipment for Your Application, Kaeser Compressors de Guatemala y Cia. Ltda., Calzada Atanasio Tzul 21-00, Zona 12, Complejo Empresarial, El Cortijo II, Bodega 501 01012 Guatemala

8. Dr E. h. Dipl.-lng. Kurt Stoll, Festo AG & Co., Esslingen, 'From the first applications of compressed air and its subsequent utilisation to the technical systems of today', 1997

9. Energy, SAVINGS in Compressed Air Systems, Kaeser Compressors, Inc. P.O. Box 946, Fredericksburg, VA 22404, www.kaeser.com

10. H Meixner & R Kobler, Maintenance of pneumatic equipment and systems, Festo Didactic, 1st Edition, 1977

11. H. Meixner & R. Kobler, Introduction to pneumatics, 2nd Edition, Festo Didactic, 1977

12. Hesse, Examples of pneumatic applications, Blue Digest for Automation, Festo, 1999

13. Hesse, Grippers and their applications, Blue Digest for Automation, Festo AG & Co, 1998

14. J P Hasebrink, R Kobler, Fundamentals of pneumatic control engineering, 3rd Edition Festo Didactic, 1989

15. Joji P., Pneumatic controls, Wiley India Pvt Ltd, New Delhi, 2008

16. Operating conditions and standards in pneumatics, FESTO

17. Pneumatics – 2000, Norgren product catalogue

18. SIMPLIFIED VALVE CIRCUIT GUIDE A GUIDE TO UNDERSTANDING PNEUMATIC DIRECTIONAL CONTROL VALVES, NORGREN

19. The standard ISO 8573-1, Compressed air – Part 1: Contaminants and purity classes

Fluid Power Educational Series Books

1. Pneumatic Systems and Circuits -Basic Level (In the SI Units)
2. Industrial Pneumatics -Basic Level (In the English Units)
3. Pneumatic Systems and Circuits -Advanced Level
4. Electro-Pneumatics and Automation
5. Design of Pneumatic Systems (In the SI Units)
6. Design Concepts in Pneumatic Systems (In the English Units)
7. Maintenance, Troubleshooting, and Safety in Pneumatic Systems
8. Industrial Hydraulic Systems and Circuits -Basic Level (In the SI Units)
9. Industrial Hydraulics -Basic Level (In the English Units)
10. Hydraulic Fluids
11. Hydraulic Filters: Construction, Installation Locations, and Specifications
12. Hydraulic Power Packs (In the SI Units)
13. Power Packs in Hydraulic Systems (In the English Units)
14. Hydraulic Cylinders (In the SI Units)
15. Hydraulic Linear Actuators (In the English Units)
16. Hydraulic Motors (In the SI Units)
17. Hydraulic Rotary Actuators (In the English Units)
18. Hydraulic Accumulators and Circuits (In the SI Units)
19. Accumulators in Hydraulic Systems (In the English Units)
20. Hydraulic Pipes, Tubes, and Hoses (In the SI Units)
21. Pipes, Tubes, and Hoses in Hydraulic Systems (In the English Units)
22. Design of Industrial Hydraulic Systems (In the SI Units)
23. Design Concepts in Industrial Hydraulic Systems (In the English Units)
24. Maintenance, Troubleshooting, and Safety in Hydraulic Systems
25. Hydrostatic Transmissions (HSTs) (In the SI Units)
26. Concepts of Hydrostatic Transmissions (In the English Units)
27. Load Sensing Hydraulic Systems (In the SI Units)
28. Concepts of Load Sensing Hydraulic Systems (In the English Units)
29. Electro-hydraulic Proportional Valves
30. Electro-hydraulic Servo Valves
31. Cartridge Valves
32. Electro-hydraulic Systems and Relay Circuits
33. Practical Book: Pneumatics - Basic Level
34. Practical Book: Electro-pneumatics - Basic Level
35. Practical Book: Industrial Hydraulics – Basic Level
36. Programmable Logic Controllers and Programming Concepts
37. Compressed Air Dryers
38. Hydraulic Circuits – Identification of Components and Analysis

For more details, please visit: **https://jojibooks.com**

Made in the USA
Columbia, SC
10 July 2024